George Viner Ellis, George Henry Ford

Illustrations of Dissections

George Viner Ellis, George Henry Ford

Illustrations of Dissections

ISBN/EAN: 9783337366735

Printed in Europe, USA, Canada, Australia, Japan

Cover: Foto ©berggeist007 / pixelio.de

More available books at **www.hansebooks.com**

LUSTRATIONS OF DISSECTIONS

IN A

SERIES OF ORIGINAL COLORED PLATES

THE SIZE OF LIFE

REPRESENTING THE

DISSECTION OF THE HUMAN BODY.

BY

GEORGE VINER ELLIS

PROFESSOR OF ANATOMY IN UNIVERSITY COLLEGE, LONDON

AND

G. H. FORD, Esq.

E DRAWINGS ARE FROM NATURE BY MR. FORD, FROM
BY PROFESSOR EL

*(Reduced on a uniform scale, a~~ ~roduced in facsimile, expressly for
Standard Medical Authors.)*

VOL. I

SECOND EDITION

NEW YORK
WILLIAM WOOD & COMPANY
1882

PREFACE.

This Volume contains a concise description of a series of Anatomical Plates in folio in a separate Atlas (in this edition the Plates are bound with the text.—Am. Publishers), with some remarks on the practical applications of Anatomical facts to Surgery. The purpose for which the Plates are designed, and the circumstances connected with their production are explained below.

With the view of carrying out the pictorial representation of dissections, the part of the Human Body to be illustrated is divided into suitable stages or regions; and the muscles, bloodvessels, and nerves of each region are shown in layers in the natural order of succession, so that their mutual connections may be brought before the eye at one and the same time.

The Illustrations comprise views of the Head and Neck, the upper Limb, the Perinæum, the Abdominal parietes, the Pelvis, and the lower Limb. All the Figures are drawn of life-size from actual dissections; and they are printed in colors with the object of making them as true pictures as possible of Nature, and more serviceable as copies for the student to imitate. Only such dissections were prepared for the Drawings as may be commonly seen in the practical Anatomy Room; and the minute detail, whose counterpart the student with average manual dexterity could not produce without some difficulty and loss of time, was intentionally omitted. Delineations of the ligaments, the viscera of the cavities of the Body, and the organs of the Senses, are not included in the Plates now published.

The labor connected with this Work was divided between its two authors, that part being apportioned to each which he was best fitted by previous knowledge and experience to execute. To Mr. Ford were assigned the original Drawings, and the chromo-lithography; and to him is due the merit of portraying with so much effect and exactness the natural appearance of the parts dissected. Upon me

rests the responsibility of the selection of the Illustrations, the
of the dissections, and the accuracy of the whole.

During the progress of this undertaking, which was continued through several years, other engagements necessitated my having recourse occasionally to the senior Students of the College for the help of their hands. To those Students, and to Mr. Samuel Onley in particular, I gladly offer my thanks for their assistance. And to the late Mr. J. S. Cluff, Demonstrator of Anatomy, I am greatly indebted for the valuable aid he afforded me at all times.

Before closing this retrospect of the task now finished, I may advert to the difficulties attendant on the printing in colors of such complicated Figures, and to the successful way in which they were overcome by **Mr. West**.

GEORGE VINER ELLIS.

UNIV. COLL. LOND.,
May 1*st*, 1876.

CONTENTS.

THE UPPER LIMB.

	PAGE
PLATE I.—THE SUPERFICIAL MUSCLES OF THE THORAX, AND THE AXILLA WITH ITS CONTENTS	1
Superficial thoracic muscles	1
Boundaries of the axilla	3
Arteries of the axilla	5
Veins of the axilla	9
Nerves in the axilla	10
Nerves of the brachial plexus	10
Lateral cutaneous nerves of the thorax	11
Lymphatics of the axilla	12
Fat in the axilla	13
PLATE II.—THE AXILLARY VESSELS, AND BRACHIAL PLEXUS OF NERVES, WITH THEIR BRANCHES	14
Superficial prominences of bone	14
Deep muscles of the thorax	16
Axillary artery and branches	19
vein and branches	24
Brachial plexus, and its branches	25
Lymphatic glands	26
PLATE III.—SUPERFICIAL VEINS AND NERVES IN FRONT OF THE BEND OF THE ELBOW	27
Fascia of the limb, and the biceps	27
Cutaneous veins and blood-letting, with injury of the brachial artery	28
Brachial artery, with ligature of	34
Cutaneous nerves at the elbow	36
PLATE IV.—SUPERFICIAL VIEW OF THE ARM ON THE INNER SIDE, WITH THE PARTS UNDISTURBED	37
Surface-marking of the arm	37
Fascia and muscles	38
Veins of the arm	40
Brachial artery, its ligature, and branches	41
Nerves of the arm	46
Lymphatics of the arm	47

CONTENTS.

	PAGE
PLATE V.—THE SHOULDER, AND THE MUSCLES AT THE BACK OF THE SCAPULA	48
Scapula and shoulder muscles	48
Arteries of the shoulder	52
Nerve of the shoulder	53
PLATE VI.—THE TRICEPS MUSCLE BEHIND THE HUMERUS, AND SOME SHOULDER MUSCLES	54
Triceps extensor muscle, and fracture of the olecranon	54
Arteries of the arm and shoulder	56
PLATE VII.—THE MUSCULO-SPIRAL NERVE IN THE ARM, AND THE PROFUNDA VESSELS	58
Muscles of the arm and shoulder	58
Vessels at the back of the arm	60
Nerves at the back of the arm	61
PLATE VIII.—SURFACE VIEW OF THE FOREARM, WITH THE PARTS UNDISTURBED	63
Superficial muscles	63
Hollow before the elbow	66
Radial artery and branches : ligature of	67
Nerves superficial in the forearm	70
PLATE IX.—DEEP VIEW OF THE FRONT OF THE FOREARM	71
Muscles of the deep layer	71
Ulnar artery and branches: ligature of	74
Nerves of the forearm	77
PLATE X.—SUPERFICIAL AND DEEP VIEWS OF THE PALM OF THE HAND	78
Fig. i. Central muscles of the palm	79
Superficial palmar arch : wounds of	81
Superficial nerves of the hand	84
Fig. ii. Short muscles of the digits	86
Radial artery and deep palmar arch	88
Deep nerve of the hand	90
PLATE XI.—SUPERFICIAL VIEW OF THE BACK OF THE FOREARM AND HAND	91
Superficial layer of muscles	91
Radial artery at the back of the hand	96
PLATE XII.—DEEP VIEW OF THE BACK OF THE FOREARM	98
Muscles of the deep layer	98
Arteries at the back of the forearm	100
Nerve at the back of the forearm	101

THE HEAD AND NECK.

	PAGE
PLATE XIII.—BASE OF THE SKULL, AND FIRST AND SECOND VIEWS OF THE ORBIT	104
Fossæ of the base, and parts of the dura mater	104
Cranial nerves in the skull	106
Vessels of the base of the skull	109
Highest muscles of the orbit, and the lachrymal gland	111
Vessels of the orbit	112
Superficial nerves of the orbit	114
PLATE XIV.—SINUSES OF THE DURA MATER, AND TWO DEEP VIEWS OF THE ORBIT	115
Sinuses of the dura mater	116
Deep muscles of the orbit	117
Nerves deep in the orbit	118
PLATE XV.—THE ANATOMY OF THE SIDE OF THE NECK BEHIND THE STERNO-MASTOID MUSCLE	121
Lateral muscles of the neck	121
Posterior triangular space—its boundaries and contents	124
Arteries in the space, and ligature of the third part of the subclavian	127
External jugular vein and blood-letting	131
Nerves in the triangular space	133
PLATE XVI.—SURFACE VIEW OF THE NECK IN FRONT OF THE STERNO-MASTOIDEUS MUSCLE	135
Surface objects visible without displacement of any part	135
Sterno-mastoideus and the fascia	137
Connections of the salivary glands	138
Superficial arteries of the neck	139
veins of the neck	140
Cutaneous nerves of the neck	140
PLATE XVII.—VIEW OF THE FRONT OF THE NECK AFTER DISPLACEMENT OF THE STERNO-MASTOIDEUS	141
Anterior triangular space	141
Anterior muscles of the neck	143
Carotid vessels, and ligature of	146
Veins of the front of the neck	151
Nerves of the fore part of the neck	151

CONTENTS.

	PAGE
PLATE XVIII.—THE SUBCLAVIAN ARTERY AND THE SURROUNDING PARTS	153
Muscles of the subclavian region	153
Subclavian artery and branches, with ligature of the second part	154
Subclavian and deep jugular veins	156
Nerves of the subclavian region	157
PLATE XIX.—A DEEP VIEW OF THE BACK OF THE NECK	159
Deep muscles behind the spine	159
Arteries at the back of the neck	160
Nerves at the back of the neck	161
PLATE XX.—SUPERFICIAL VIEW OF THE PTERYGOID REGION	163
Muscles of mastication	163
Internal maxillary artery and branches	165
Internal maxillary and facial veins	166
Branches of inferior maxillary nerve	167
PLATE XXI.—DEEP VIEW OF THE DISSECTION OF THE PTERYGOID REGION	168
Cranial branches of internal maxillary artery	168
Inferior maxillary nerve and branches	169
PLATE XXII.—THE ANATOMY OF THE SUBMAXILLARY REGION	172
Tongue and hyoid muscles	172
Salivary glands under the jaw	174
Lingual artery and vein	174
Nerves of tongue, and the submaxillary ganglion	175
PLATE XXIII.—UPPER MAXILLARY NERVE, AND DEEP PART OF THE INTERNAL MAXILLARY ARTERY	177
Some facial muscles	177
Terminal offsets of the maxillary artery	178
Upper maxillary and facial nerves	179
PLATE XXIV.—INTERNAL CAROTID AND ASCENDING PHARYNGEAL ARTERIES, AND CRANIAL NERVES IN THE NECK	180
Deep muscles in front of the spine	181
Carotid and ascending pharyngeal arteries	182
Cranial, spinal, and sympathetic nerves in the neck	184
PLATE XXV.—EXTERNAL VIEW OF THE PHARYNX WITH ITS MUSCLES	190
The pharynx and its muscles	190
Some nerves and vessels of the larynx	192

CONTENTS.

PLATE XXVI.—INTERIOR OF THE PHARYNX, AND THE MUSCLES OF THE SOFT
    PALATE 193
        Cavity of the pharynx, and its openings 194
        The soft palate and the tonsil 197
        Muscles of the soft palate, and use of the part . . 198

PLATE XXVII.—LARYNX AND VOCAL APPARATUS, WITH THE MUSCLES, VESSELS, AND NERVES 201
Figs. ii. and iii. Cartilages of larynx, and hyoid bone . . . 201
        Articulations of laryngeal cartilages 204
        Interior of larynx, and vocal apparatus 206
    Fig. i. Muscles of the larynx governing the size of the glottis and
        the pitch of the voice 208
        Nerves of the larynx, and use 211
        Vessels of the larynx 213
        Thyroid body and the trachea 214

PLATE XXVIII.—NOSE CAVITY WITH THE BOUNDARIES AND OPENINGS INTO
    IT 215
        Cavity of the nose and its bounds . . . 216
        Spongy bones and the meatuses 218
        Mucous membrane and bloodvessels 219
        Olfactory region and nerves of the nose 220

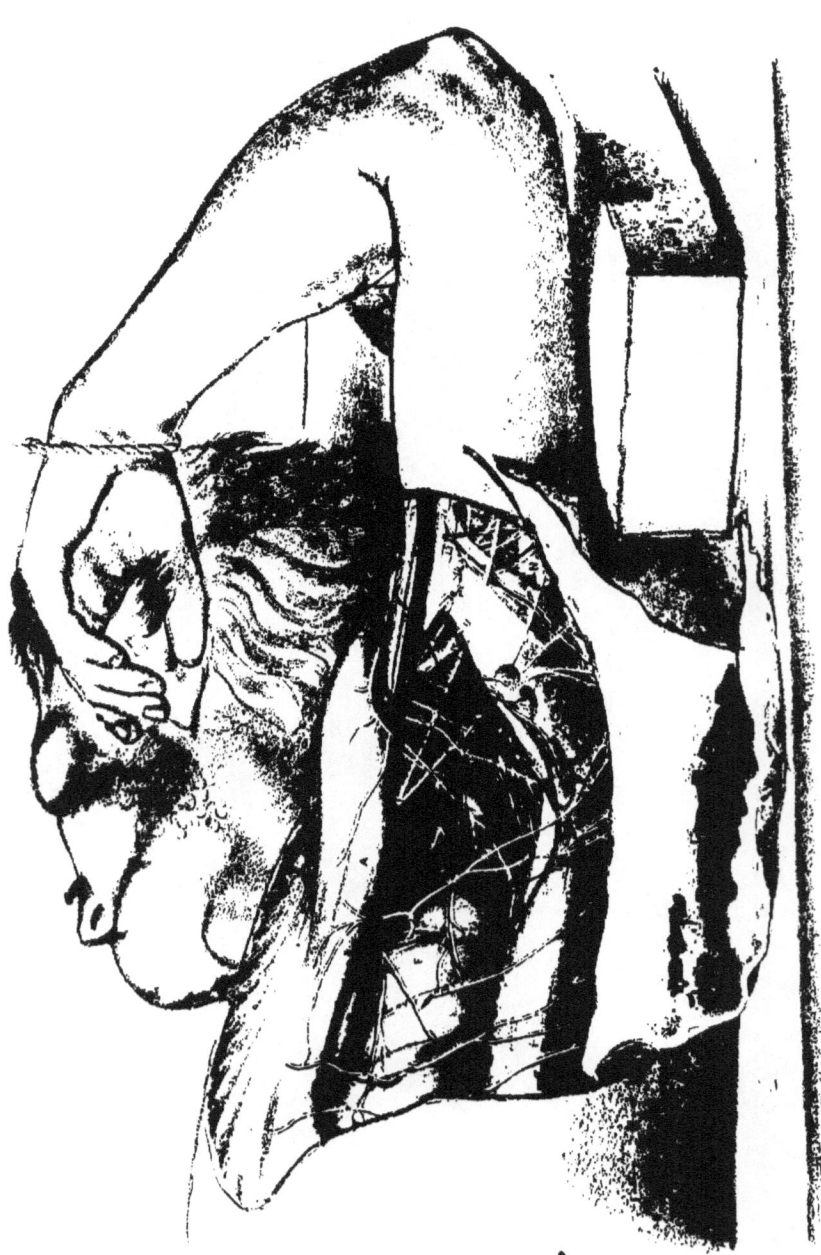

ILLUSTRATIONS OF DISSECTIONS.

DESCRIPTION OF THE PLATES.

DESCRIPTION OF PLATE I.

THE superficial muscles of the thorax, and the axilla with its contents, are delineated in this Plate. The natural size of the dissected part has been slightly reduced in the Drawing for the purpose of showing the whole of the upper limb.

In the preparation of the dissection the limb was drawn away from the trunk to render tense the muscles. Next, the integuments were divided, and the skin and fat were raised together in one large flap from the front of the thorax and the axilla by carrying the scalpel upwards and outwards from the chest to the arm, as the fibres of the muscle run, and along one muscular fasciculus at a time. Afterwards the fat was cleared out of the axilla; and the muscles bounding the space behind were laid bare in the same way as those limiting it in front.

MUSCLES OF THE THORAX AND ARM.

Two sets of muscles are displayed in the dissection; one bounding the arm-pit before and behind; and the other lying in front, and at the back of the humerus.

The former group are directed from the trunk to the limb, and move the limb forwards and backwards over the chest. Where they are fixed to the thorax they are separated widely by the ribs, but at the arm-bone they approach one another. In the interval between them, near the humerus, the large vessels and nerves of the limb are lodged.

The muscles of the arm connect the limb with the scapula, and assist the movements of the shoulder joint: they will be more fully seen in other Plates.

A. Pectoralis major.
B. Pectoralis minor.
C. Latissimus dorsi.
D. Teres major.
F. Serratus magnus.
H. Subscapularis.

K. Coraco-brachialis.
L. Biceps.
M. Fascia of the arm.
N. Triceps muscle.
P. External head of the triceps.

The *pectoralis major*, A, reaches from the chest to the arm over the front of the axilla. By its inner end (origin) it is attached to the sternum and the cartilages of the true ribs, except the last, as well as to a part of the clavicle; and it joins the tendon of the external oblique muscle of the abdomen below. By its outer end (insertion) it is fixed into the bicipital groove of the humerus. Plate ii. P.

The muscle forms the main part of the anterior boundary of the axilla; and towards its lower end near the arm-pit the mamma or breast rests on it.

Should the breast be diseased so as to render necessary its removal, the limb should be placed during the operation in the position shown in the Illustration, and the scalpel should be carried in the direction of the fibres in detaching the mass to be extirpated.

The *pectoralis minor*, B, is extended, like the preceding, from the chest to the limb in front of the axilla. Only a very small part is now visible: for a view of the muscle see Plate ii. B.

The *latissimus dorsi*, C, resembles in its position behind the axilla the large pectoral muscle in front. Arising below from the spinal column, the pelvis, and the lower ribs, it is inserted into the bicipital groove of the humerus. The upper edge of the muscle has been everted in the Drawing to bring into sight the vessels and the nerve lying inside it.

Oftentimes a fleshy slip is continued from the latissimus over the axillary vessels to join the pectoralis major, the fascia of the arm, or the coraco-brachialis muscle.

The *teres major*, D, lies behind the latissimus, by which it is partly concealed. Attached below to the scapula, it is inserted into the humerus beneath the broad muscle of the back, C.

The three muscles, pectoralis major, latissimus, and teres, converging from the trunk and scapula to the upper part of the arm-bone, will approximate the raised movable limb to the trunk when they act simultaneously; and the limb will be moved forwards or backwards in accordance with the preponderating power of the pectoral or of the two others. Their action may occasion dislocation of the humerus under the following circumstances. In falling, with the arm outstretched, the elbow comes into contact with the ground, and renders the lower end of the humerus immovable; and if the muscles then act suddenly and forcibly they will draw down the upper end of the bone, which is free to move, and bring it into the axilla.

The *serratus magnus*, F, shuts out the ribs from the axilla, as it is directed backwards from the chest to the scapula. Its separate slips of origin from the ribs (serrations) are marked by the passage between them of nerves, and here and there of small vessels.

The *subscapularis*, H, fills the hollow of the scapula and excludes this bone from the axilla. The upward and outward direction of its fibres over the shoulder joint is indicated in the Figure. For a description of the muscle, see the explanation of Plate ii.

The *arm muscles* are the coraco-brachialis, K, and biceps, L, in front of the humerus, and the triceps, N, behind that bone. The share taken by the two last muscles in the outline of the limb may be seen in the Drawing: their anatomy will be studied in other dissections.

The *coraco-brachialis*, K, attached as the name expresses, is dis fully in Plate ii. As here seen, it is partly subcutaneous, and lies beneath the pectoralis major. The swell of the muscle is felt through the teguments, and serves as the surgeon's guide to the a artery along its inner edge.

BOUNDARIES OF THE AXILLA.

The axilla corresponds with the surface depression of the arm-pit. As it is a portion of the space included by the thoracic muscles converging to the humerus, it has necessarily a pyramidal form; and it is limited in front and behind chiefly by those muscles, whilst inside it is the chest, and on the outside the humerus. This intermuscular interval lodges the large vessels and nerves of the limb, with lymphatics, and contains a loose granular fat.

Along the fore part lies the pectoralis major, A, reaching from apex to base, and forming by its lower or free edge the anterior fold of the ; and underneath it, constructing only a small part of this bound- the pectoralis minor, B. After the fat had been removed from the space, the pectoralis sank down somewhat, as the Drawing indicates, in consequence of the body not being very fresh.

Bounding the interval behind are three muscles, viz., the subscapularis, H, the latissimus, C, and the teres major, D: the two latter muscles enter into the formation of the lower part of this boundary, and extend lower down the limb than the pectoralis. Within the edge of the latissimus, here shown everted, is a hollow containing vessels, nerves, and glands, in which pus may burrow, or slightly enlarged glands may lie undetected by the fingers.

On the inner convex side of the axilla is the serratus magnus, F, covering the four highest ribs and their intercostal muscles.

On the outer side, where the space is limited, are placed the humerus and the contiguous part of the scapula, with the coraco-brachialis, K, and biceps, L, muscles.

The base, or the elongated lower opening of the axilla, is wider at the chest than at the arm; but it is not so wide before as after the dissection, because the parts when cleaned separate from each other. A rather dense fascia closes the axillary space in this direction, and impedes the advance of pus to the surface.

The apex or narrowed part of the space joins the root of the neck, and lies between the chest and the scapular arch. It cannot be recognized in this Figure, but it may be observed more completely in Plate ii.

The state of the muscular boundaries of the axilla is much altered by the position of the limb; for the tenseness is diminished when the arm is adducted, and increased when the arm is abducted from the thorax. And the depth will vary, in like manner, with change in the position of the limb. These facts may be remembered with advantage in any endeavor to estimate the size of the tumor in the axilla.

In consequence of the position of this intermuscular space at the inner side of the shoulder bone, and of the loose nature of the fatty tissue contained in it, the movements of the scapula over the chest are facilitated. And from its situation on that side of the shoulder joint to which flexion is made, the large vessels and nerves of the limb are transmitted through it beyond the joint without injury from stretching in the motions of that articulation. A corresponding hollow exists in the lower limb in front of the hip-joint.

In the undissected limb the student may recognize by the eye and the touch the prominence of the cords of the large axillary nerves and vessels along the side of the arm; and if the arm is raised and freely moved by one hand, whilst the two fore fingers of the other are pressed into the arm-pit, the moving head of the humerus may be felt through the skin and fascia. During life the hollow is useful to the surgeon in his attempts to discover the nature and size of an enlargement, such as an aneurism, in this situation, or the position of the dislocated head of the humerus. But the size of the space interferes with the detection of small tumors, like an enlarged gland; for these may extend upwards and inwards towards the chest in the loose fat, and acquire considerable size before their presence will be indicated by any external swelling.

ARTERIES OF THE AXILLA.

The lower end of the axillary artery with its branches are now brought under notice; but only a small narrow strip of the arterial trunk can be seen in the Figure in consequence of its connections with veins and nerves being preserved. This part of the vessel is uncovered by muscle, and is in contact with the common investing parts of the limb. A superficial wound of the limb may lay it open. It supplies branches to the chest and the shoulder.

- *a*. Axillary artery.
- *b*. Long thoracic.
- *d*. Alar thoracic.
- *e*. External mammary.
- *f*. Subscapular.
- *g*. Dorsal scapular.
- *h*. Posterior circumflex.
- *n*. Cutaneous artery with the internal cutaneous of the musculo-spiral nerve.
- *o*. Artery to the long head of the triceps muscle.

The *axillary* or main artery of the upper limb crosses the space from the chest to the arm beneath the pectoral muscles, but only the part between the lower border of the pectoralis major, A, and the lower border of the teres major, D, is delineated. Its position is marked by the swell of the coraco-brachialis muscle, K; and its depth from the surface is very slight, only the tegumentary coverings of the limb concealing it, so that it can be readily reached in an operation, or it can be compressed with ease against the humerus.

Its connections with the muscles around are the following:—Behind are the subscapularis, H, the latissimus, C, and teres major, D; and on the outer side is the coraco-brachialis, K.

Its companion vein (axillary) is placed on the inner or chest side, partly concealing the artery, and has to be drawn aside in attempts to tie the arterial trunk.

Large nervous cords lie around the artery:—outside is the median nerve, 22; inside the ulnar, 21, and the nerve of Wrisberg, 11 (which is sometimes nearer the artery). Superficial to the vessel is the large internal cutaneous, 18; and deeper than it or beneath, is the musculo-spiral, 13 (which is drawn somewhat inwards below); another nerve beneath the artery for a short distance is the circumflex, 12.

Two named branches, the subscapular, *f*, and the posterior circumflex, *h*, leave this part of the artery opposite the edge of the subscapularis muscle.

Ligature of the vessel. The origin of several branches from the axillary artery opposite the shoulder joint, would interfere with the application of a ligature at that spot; but for a distance of two inches beyond (towards the arm) the vessel is free from any large branch, and might be tied without risk of hæmorrhage.

Suitable as the lower part of the axillary trunk seems to be for an operation at this place, surgeons have not chosen it for the application of a ligature. Doubtless the advantages offered by tying the brachial and subclavian arteries have caused those vessels to be selected in preference to the axillary; but the small channels that remain for the collateral circulation after the main artery is secured may have had some influence in inducing surgeons to let the operation on this part of the axillary artery remain unpractised. For the only collateral vessels to carry on the circulation after the occlusion of the trunk would be the small and indirect anastomoses through the following muscles, viz., the coraco-brachialis, biceps, and long head of the triceps; and through the shaft of the humerus.

Although the collateral vessels are so small and indirect, they are sufficient for carrying on the circulation, as the artery has been tied with success in this situation. M. Blandin secured the vessel here in a man who received an injury of the artery from the discharge of a gun. The man recovered.*

Should the surgeon be called upon to tie the artery he should keep in mind its situation along the edge of the coraco-brachialis, with its companion vein on the thoracic side and partly concealing it, and with large

---

* Traité d'Anatomie topographique, p. 506 : Paris, 1834.

nerves of the brachial plexus around it. Though the vessel is usually very superficial, it may be placed under muscular fibres directed over it from the latissimus to the pectoralis.

Two other points deserve attention also : Firstly, that two arteries instead of one may be found as often as 1 in 10.* Secondly, that not unfrequently the nerves, which should serve as the deep guide to the artery, are not placed around the parent trunk, but encircle a large branch formed by the conjunction of the usual offsets at this spot with some of the branches which are derived, as a rule, from the brachial artery.

When practising on the dead body the tying of the part of the artery issuing from the axilla the following directions may be observed :—

The limb is to be placed at right angles to the trunk, and the operator stands between the two.

With the eye fixed on the prominence of the coraco-brachialis muscle, K, which is the superficial guide to the vessel, a cut is to be begun in the hollow of the arm-pit, and to be continued along the side of the muscle for two inches ; but the knife is not to be carried deeper at first than through the skin and fat.

The axillary vein will now be recognized through the deep fascia or aponeurosis of the limb by its blue color ; and the aponeurosis being divided along the outer edge of the vein as far as the cut in the integuments, this vessel may be detached with care from the subjacent parts, and drawn inwards with a narrow retractor.

Next, the brachial plexus around the artery will serve as the deep guide. Search is to be made for the axillary trunk by cutting a piece of fat from the hollow out of which the vein has been drawn ; and it is to be made in a horizontal direction or towards the humerus, instead of backwards towards the axillary fold. The operator tries to find the artery in the midst of the nerve-trunks without attempting to distinguish the individual nerves.

After the thin arterial sheath has been opened by the knife, the needle may be passed readily, and the vessel is to be tied with as little displacement as possible.

When the artery is not surrounded by the nerves of the brachial plexus,

---

* The facts on which this statement rests will be found in the Surgical Anatomy of the Arteries of the Human Body, by Richard Quain, F.R.S. : London, 1844.

as referred to above, it will be nearer the vein and the surface than it is usually.

In the treatment of aneurism of the lower end of the axillary trunk Professor Syme has recently recommended a return to the old practice of opening the sac, and after removing the contents, putting a thread around the vessel above and below the sac. In a postscript to a communication published in the Medico-Chirurgical Transactions* he says:— "On the 15th of August, in accordance with the principles above explained, I performed the old operation for aneurism, not traumatic, at the lower part of the axilla, in a gentleman about fifty, recommended to my care by Dr. Embleton, of Embleton in Northumberland. The patient returned home on the 5th of September."

Branches of the artery. The lower part of the axillary artery supplies the following branches to the wall of the chest and the shoulder.

Long thoracic artery, b, lies in the axilla along the angle formed by the meeting of the anterior and inner boundaries, and may be injured by a cut made along the anterior axillary fold. Its origin is concealed by the pectoralis major.

Alar thoracic branch, d. This small artery to the arm-pit was present in this body, though it is generally absent (Quain). It is distributed to some of the glands, and to the fat of the axilla.

The *subscapular artery, f,* is the largest offset, and arises opposite the edge of the subscapularis muscle. Taking the border of the muscle as its guide, it reaches the chest, to which and the shoulder it is distributed. A companion vein and nerve run with it, and all are secured from external injury by the projecting margin of the latissimus dorsi. Many offsets enter glands and the continguous muscles: and one, *g,* larger than the rest, is the dorsal scapular artery.

The *posterior circumflex artery, h,* arises close beyond the preceding, and winds behind the humerus to the shoulder with the nerve, 12, of the same name. Its distribution is represented in Plate v.

An *anterior circumflex artery,* not now visible, crosses between the humerus and the coraco-brachialis to the shoulder.

The *external mammary, e,* is a long slender irregular branch, which is directed across the axilla to the chest, lying about midway between the

---

* On the Treatment of Axillary Aneurism, by James Syme, F.R.S. Edin., vol. 43, p. 143 : London, 1860.

anterior and posterior folds. It supplies the glands, and the wall of the chest, assisting the long thoracic artery.

Muscular and cutaneous branches.—Small offsets near the end of the artery supply the coraco-brachialis, K, and the long head of the triceps, N. And a cutaneous twig, *n*, accompanies the internal cutaneous branch of the musculo-spiral nerve.

An inspection of the Drawing will suffice for showing the vessels or nerves likely to be injured in wounds into the axilla, or in incisions made into it by the surgeon. Along the anterior boundary, where this joins the chest, are placed the long thoracic vessels; and lying along the posterior boundary, but within the margin of the latissimus, are the subscapular vessels and nerves with glands. On the side of the limb, or at the outer part, the trunks of the axillary vessels and nerves, and the cords of the brachial plexus are aggregated together; whilst on the side of the chest there is only an occasional small artery. If an incision is to be made into the arm-pit the surgeon should select the inner boundary as the freest from vessels, and should direct the knife about midway between the anterior and posterior folds.

VEINS OF THE AXILLA.

Only the position of the chief vein to the artery was retained in the dissection; and the smaller veins, which would complicate the drawing without corresponding utility, were removed.

l. Axillary vein.	*p*. External mammary vein.
m. Subscapular vein.	*r*. Cutaneous and muscular vein.

The *axillary vein*, *l*, the chief trunk of the limb, is continuous in the arm, just beyond the axilla, with the cutaneous vein—basilic. Placed on the inner or thoracic side of, and partly concealing the axillary artery, it receives small contributing veins corresponding with the arterial branches. Some of these are seen in the Plate.

Frequently two veins instead of one are present in the lower part of the axillary space.

Through this vein nearly the whole of the blood of the limb below the shoulder is conveyed onwards; and interruption to its current will occasion congestion in the parts to which its roots extend. A tolerably complete occlusion of this main circulating channel, as in the case of a slowly-growing tumor, will not only give rise to congestion, but will cause serous fluid to transude through the coats of the vessels into the surrounding textures.

NERVES OF THE AXILLA.

The nerves in this dissection are derived from two sources :—Those on the side of the chest are offsets of the intercostal nerves, and appear between the digitations of the serratus magnus muscle; and those lying around or near the axillary vessels belong to the brachial plexus.

LATERAL CUTANEOUS OF THE THORAX.

1. Lateral cutaneous branch of the second intercostal nerve (the highest of the set).
2. Offset of third intercostal.
3. Offset of fourth intercostal.
4. Offset of fifth intercostal.
5. Offset of sixth intercostal.
6. Anterior branch of the offset of the second intercostal.
***Anterior branches of the offsets of the other intercostal nerves.
7. Communicating branch to the nerve of Wrisberg from the offset of the second intercostal nerve.

BRACHIAL PLEXUS.

8. Nerve to the teres muscle.
9. Subscapular nerve.
10. Nerve to the serratus magnus.
11. Nerve of Wrisberg.
12. Circumflex nerve.
13. Musculo-spiral.
14. Offset of the musculo-spiral to the triceps.
16. Internal cutaneous of the musculo-spiral.
18. Large internal cutaneous.
20. Offset of the internal cutaneous to the integuments.
21. Ulnar nerve.
22. Median nerve.

NERVES OF THE BRACHIAL PLEXUS.

The *median*, 22, is the companion nerve to the axillary artery, and is placed on the outer side.

The *ulnar nerve*, 21, smaller than the median, but without branch like it, lies to the inner side of the arterial trunk.

The *musculo-spiral*, 13, occupies, naturally, a position beneath the vessel, but it has been pulled inwards, and is represented in the Plate as on the inner side. Here it furnishes two small branches: one muscular, 14, to the inner and middle heads of the triceps; the other is the internal cutaneous, 16, which is distributed to the integuments of the back of the arm.

The *circumflex nerve*, 12, accompanies the artery of the same name, *h*, to the deltoid muscle. See Plate v.

Large internal cutaneous nerve, 18, lies on the axillary artery, and gives a small cutaneous offset, 20, to the integuments of the arm over the situation of the bloodvessels; but its direction has been altered by the displacement of the skin.

Small internal cutaneous nerve, 11, (nerve of Wrisberg) issues beneath, though sometimes through the axillary vein, and is joined by a branch 7, from the highest lateral cutaneous nerve of the thorax. Its position close to the vein has been disturbed by the dragging of the skin.

Muscular branches. The nerve to the teres major, 8, and the nerve to the latissimus, 9, are directed with the subscapular vessels along the back of the arm-pit to their destination; the former gives an offset to the subscapularis muscle.

The nerve to the serratus magnus is continued on the surface nearly to the lower border of its muscle, giving backwards offsets to the fleshy fibres. Its origin is connected with the trunks of the fifth and sixth cervical nerves in the neck.

Pressure applied to the nerves of the brachial plexus may occasion pain, or loss of power and feeling, according to its degree, in a greater or smaller part of the limb. In the use of crutches the weight of the body acts injuriously on the nerves, for the arm is arched over the top of the crutch, and the nerves are compressed between the humerus and the artificial prop of the body. This inconvenience may be remedied by the crutch-head being so constructed as to bear least on the centre of the arm over the large nerves.

LATERAL CUTANEOUS NERVES OF THE THORAX.

Five *lateral cutaneous branches* of the intercostal trunks were laid bare in the dissection; they appear lax after they have been separated from the surrounding fat. The branches directed forwards over the pectoralis were necessarily detached from the skin, and were then laid on the surface of the muscle.

As the first intercostal trunk does not furnish commonly any lateral cutaneous branch, the nerves shown are derived from the five next intercostal trunks. Each branch divides into two parts (anterior and posterior) as it issues between the ribs, and these terminate on the lateral part of the thorax.

The anterior offsets, 6, * * *, end in the integuments covering the pectoralis major; and the posterior, 1, 2, 3, 4, 5, somewhat larger in size, ramify in the skin of the arm, and in that over the latissimus dorsi. In this body the third nerve wanted an anterior offset.

The highest and largest of the lateral cutaneous nerves,—that from the second intercostal trunk, differs in some respects from the others. Its anterior branch, 6 (laid on the pectoralis, and not always present), supplies the arm-pit as well as the teguments on the pectoralis major· its posterior branch, 1, called intercosto-humeral, reaches the integuments of the back of the arm, and gives a communicating offset, 7, to the nerve of Wrisberg.

LYMPHATICS OF THE AXILLA.

Only a few of the glands of the axilla were retained in the dissection, and these have fallen, necessarily, from their natural position after the removal of the fat in which they are imbedded.

† † † Anterior group of the axillary glands.
s s s. Posterior group of glands.

t. One of the group of glands along the side of the axillary vessels.

About ten or twelve in number, the glands vary much in their shape and size. They have the following general linear arrangement in sets. The greater or hinder group lies along the subscapular vessels within the edge of the latissimus dorsi; but after the dissection of the axilla they hang in front of the muscle by their small vessels, as is shown in the Plate. Another or anterior group is nearer the fore part of the axilla, in connection with the long thoracic and external mammary arteries. And a third set is placed along the large axillary vessels.

Each collection of glands has for the most part its own set of lymphatic vessels. Thus the anterior group receives lymphatics from the fore part of the thorax and from the mamma: the posterior group is oined by the lymphatics from the side of the chest, and from the back; and that along the bloodvessels transmits lympathics from the upper limb. The lymphatic vessels, after passing through their respective glands, unite into one or more trunks at the top of the axilla, and open into the lymphatic duct of the same side.

Disease in the part from which the lymphatic vessels are derived may occasion enlargement of the group of glands through which those vessels are transmitted; and the knowledge of the destination of the lymphatics will suggest the glandular group likely to be affected:—Thus, a poisoned wound of the hand, as in dissection, will cause inflammation of the glands by the side of the axillary vessels; and so forth.

In making the necessary examination to detect disease of the glands, the limb should be approximated to the side to relax the muscles and fascia bounding the axilla, and thus to permit easier and freer manipulation. The glands near the axillary vessels follow the arm when this is elevated.

Enlargement of a gland may surround or press upon the intercostohumeral nerve, or the nerve of Wrisberg, and occasion numbness in the part to which either nerve is distributed.

Should extirpation of a diseased gland be considered advisable, the surgeon should be mindful that it has large bloodvessels, in the form of a foot-stalk, which are derived from the contiguous vessels; and he should secure the vascular pedicle with a thread before he cuts it through. If this precaution is neglected the divided vessels retract into the loose areolar tissue of the axilla, and may continue to bleed at intervals so as to endanger life.

FAT IN THE AXILLA.

The axilla is filled with a granular fat intermixed with slight areolar tissue. Towards the apex of the space the adipose tissue diminishes. In thin bodies the quantity of the fat is less, as it is in all other parts, and the space contains a watery fluid in the meshes of the areolar tissue.

The presence of fat favors in this space, as elsewhere, the accumulation of pus, which burrows amongst the loose fatty material instead of making its way to the surface through the intervening fascia. Much inconvenience and suffering may be avoided by an early incision for the escape of the confined pus.

DESCRIPTION OF PLATE II.

The Figure represents the deep dissection of the front of the chest, and that of the axillary vessels and the brachial plexus of nerves with their branches.

The dissection is to be made by cutting through and reflecting the pectoralis major. To render tense and distinct the sheath of the axillary vessels, place the limb at right angles to the trunk, and rotating it inwards, press it backwards, so as to raise the clavicle from the chest. Unless this position of the arm is kept, the loose costo-coracoid sheath may be removed with the fat.

SUPERFICIAL PROMINENCES OF BONE.

At the upper part of the region dissected is the bony loop of the scapular arch, which is formed by the clavicle, J, and the scapula, and separates the neck from the chest and the limb. It serves the purpose of articulating the upper limb, and furnishes points of attachment to muscles moving the humerus. Injury of the arch, sufficient to break it, will arrest the free movements of the shoulder joint, and interfere with the action of the muscles.

Part of the arch is subcutaneous, and the forefinger when carried along it traces successively the outline of the clavicle, acromion, and spine of the scapula. From its slight depth injuries of it are easily ascertained, because all irregularity of the surface can be detected at once with the finger.

On the inner side of the shoulder joint below the clavicle, and projecting at the edge of the deltoid muscle, R, is the coracoid process. It gives attachment to the three muscles B, K, and L, as well as to a strong ligament (coraco-clavicular), which passes from its upper and hinder part to the under surface of the clavicle, and unites together firmly the two bones. On the surface of the body this projecting osseous point can be felt between the deltoid and pectoral muscles.

In consequence of the clavicle acting as a prop to keep the shoulder from the trunk, it is very liable to be broken. By direct violence it may be shattered at any spot; but force applied to the outer end through a fall or a blow produces fracture generally about the middle of the bone.

In fracture of the shaft, that is, internal to the line of the coracoid process and the strong ligament joining this part to the clavicle, the scapula and shoulder joint, having lost their support, fall downwards and inwards towards the chest, forcing the outer past the inner fragment; and the large muscles of the chest which are inserted into the humerus assist in bringing the shoulder into closer apposition with the thorax. The inner fragment, freed from the weight of the shoulder, remains in its natural position, though it appears more than usually prominent; and the muscles attached on opposite sides, viz., the great pectoral and sterno-mastoid, may act also as antagonists, and prevent its displacement.

If the fracture takes place opposite the strong ligament uniting the coracoid process with the clavicle, the scapula remains attached to the clavicle by that ligament, though not perfectly supported by it, and the shoulder falls but little towards the chest.

In fracture external to the ligament, there is, however, considerable displacement of the bone, for the outer detached end being loose, and being acted on by the trapezius muscle, is placed in front of, and may take even a position at a right angle to the other.*

In replacing the external fragment of a broken shaft of the clavicle, the piece of bone must be moved outwards indirectly by forcing outwards the scapula; and it is to be raised to the level of the inner fragment by lifting and supporting the elbow.

At the outer part of the dissection is the projection of the shoulder, which is produced by the upper end of the arm bone covered by the deltoid muscle, R. When the limb is pendent the swell of the muscle runs into that of the arch formed by the clavicle and acromion; and when the limb is raised and lowered, the arm bone can be felt moving under the muscle.

In dislocation of the shoulder joint the upper end of the humerus sinks down from the deltoid; and a hollow then occupies the site of the prominence. This injury is accompanied necessarily by unnatural direc-

---

* A Treatise on Fractures in the Vicinity of Joints, by Robert William Smith, M.D.: Dublin, 1850; p. 210.

tion of the shaft of the arm bone forwards or backwards, and by a sharp edge along the bony arch of the clavicle and the acromion process.

MUSCLES OF THE THORAX AND ARM.

A. Pectoralis major.
B. Pectoralis minor.
C. Latissimus dorsi.
D. Teres major.
F. Serratus magnus.
H. Subscapularis.
J. Clavicle with the cut attachment of the pectoralis major.

K. Coraco-brachialis.
L. Biceps, its short head.
O. Biceps, the long head.
N. Triceps extensor brachii.
P. Insertion of pectoralis major.
R. Deltoid muscle.
S. Subclavius muscle.
V. Costo-coracoid membrane.

The muscles of the chest and shoulder, which are partly displayed in the Drawing, give to the scapula and the shoulder joint some of their varied movements.

The scapula has a gliding motion over the ribs, and can be moved in opposite directions. It is drawn forwards by the small pectoral, B, and serratus magnus muscle, F, which attach it to the chest.

The shoulder being a ball and socket joint is provided with muscles on opposite sides; but only two are now evident, viz., the deltoid or great abductor, R, and the subscapularis or internal rotator, H.

In the group of thoracic muscles are included the pectoralis major and minor, the serratus magnus, the latissimus dorsi, and the subclavius.

Pectoralis major, A. After the division of the muscle the parts underneath it can be observed. It covers the pectoralis minor on the chest, and the coraco-brachialis, K, and the biceps, L and O, in the arm. Near the clavicle the subclavius muscle, S, and the costo-coracoid membrane, V, lie beneath it. Above and below the pectoralis minor the axillary vessels and nerves are covered by the great pectoral muscle alone.

At its insertion the tendon is divided into two parts, with an interval between, something like a sling. On the under piece, P, the lower chest fibres are received; and in the other (seen only in part), the upper thoracic and the clavicular fibres terminate.

The *pectoralis minor*, B, is attached to the side of the chest, and to the third, fourth, and fifth ribs; it is inserted externally into the coracoid process of the scapula, where it blends in a common tendon with the coraco-brachialis, K, and the short head of the biceps, L.

Between the chest and the shoulder the muscle forms part of the anterior boundary of the axilla, and lies over the axillary vessels and nerves; and between the muscle and the clavicle is a triangular interval —the sides being formed by that bone and the pectoralis minor, the base by the thorax, and the apex by the coracoid process—in which the upper part of the axillary artery may be tied. Its position to other vessels and nerves is so apparent as not to need farther notice.

The pectoralis minor assists the serratus, as before said, in drawing forwards the scapula; and it may act as a muscle of forced inspiration when the scapula is the fixed part.

Serratus magnus, F, covers the side of the chest, taking origin by nine fleshy slips from the eight upper ribs; and it is inserted into the base of the scapula. Its special nerve, 5, lies on the surface, and distributes offsets to it.

From the direction of its fibres the muscle is chiefly employed in moving forwards the scapula over the ribs; and, when the scapula is fixed, it will act on the ribs so as to draw them outwards, and increase the size of the chest in inspiration. It supports, too, the lower end of the scapula whilst a weight is carried on the shoulder.

Latissimus dorsi, C. The oblique direction of this muscle behind the axilla, converging with the pectoralis major to the insertion into the humerus, is more fully seen in this than in the preceding dissection. The chief notice of this muscle is given with the explanation of Plate i.

Subclavius, S. This small muscle is contained in a sheath of the costo-coracoid membrane, of which a piece has been cut away near the inner end. Named from the position to the clavicle, its origin is attached to the first rib, and its insertion is fixed into the grooved under surface of the clavicle.

It can depress the clavicle or elevate the first rib, according as the one or the other bone may be in a state to be moved.

The shoulder muscles coming into view in this dissection are, the subscapularis, teres major, and deltoid.

The *subscapularis*, II, arises from the hollowed costal surface of the scapula, and its fibres are directed outwards and upwards over the shoulder joint to their insertion into the small tuberosity and the neck of the humerus.

By its lower edge it projects much beyond the scapula, and touches the latissimus dorsi and teres major. The subscapularis supports internally the shoulder joint, of which it is one of the articular muscles.

When the arm is raised, the subscapularis assists in depressing it; and the hanging limb is rotated inwards by the muscle.

This muscle is injured in the following dislocations of the humerus. When the bone is forced into the lower part of the axilla, it may either be covered by the subscapularis; or may be driven through the muscular fibres, and, coming into contact with, press upon the mass of the axillary vessels and nerves. In the forward dislocation on the inner side of the cervix of the scapula, the head of the bone passes between the subscapularis and the scapula, separating the fleshy fibres from the blade-bone, and projects above the upper border of the muscle.*

Teres major, D. Only the general position of the teres, which extends from the lower angle and border of the scapula to the humerus, can be now seen. The muscle is described with Plate v., D.

The *deltoid muscle*, R, forms the prominence of the shoulder, and reaches from the scapular arch to the arm bone below the level of the axilla. Only the fore part of the muscle is here represented: its insertion and connections are seen in Plate v., N.

Three muscles of the arm, biceps, coraco-brachialis, and triceps, are laid bare in the dissection—the two former, which are superficial to the humerus, being much more apparent than the latter, which is behind the bone. The anatomy of the triceps will be given in the notice of Plate vi.

The *biceps* muscle consists above (origin) of two parts, long head and short head.

The short head, L, is fixed by a wide tendon to the coracoid process; and the long tendinous head, O, narrow and rounded, passes along the groove in the humerus, and through the shoulder joint, to be attached to the top of the glenoid articular surface of the scapula.

The muscle is shown lower in the arm in Plate iv.

Coraco-brachialis, K. It arises from the coracoid process of the scapula, and the tendon of the short head of the biceps; and it is inserted into the inner side of the shaft of the humerus about midway between

---

* The student will find the state of the muscles in dislocations of the shoulder joint fully treated in the article, Abnormal Conditions of the Shoulder Joint, in the *Cyclopædia of Anatomy and Physiology*, by Robert Adams, Esq., 1849.

the ends. Its upper extremity lies beneath the pectoralis major; its insertion is concealed by the brachial vessels; and the intermediate part (belly) is subcutaneous in the arm-pit, and serves as the guide to the axillary vessels. Through the fleshy fibres of the muscle the musculo-cutaneous nerve, 11, is transmitted.

If the limb is in a state of abduction, it can be brought to the side of the chest by this muscle.

The *costo-coracoid* membrane, V, is a rather strong layer of fascia between the upper limb and the neck, and is placed there apparently for the purpose of protecting the large blood-vessels. Occupying the interval between the first rib and the coracoid process, it is fixed above to the clavicle before and behind the subclavius muscle which it incases. Below it blends with the special sheath (axillary) of the blood-vessels, giving to this additional strength; and it is continued onwards beneath the small pectoral muscle, where it gradually ceases.

The *axillary sheath* around the vessels and nerves coming from the neck to the upper limb, consists in part of a prolongation from the deep fascia of the neck, and in part of a stronger layer added from the costo-coracoid membrane. It resembles the crural sheath around the blood-vessels of the lower limb, and is funnel-shaped like that tube. In it are the axillary artery and vein, and the brachial plexus; and piercing the front are branches of those trunks, viz., the cephalic vein, *l*, the acromial thoracic artery, *c*, and anterior thoracic nerves, 2, 3. In a dissection of the axillary sheath the tube is to be opened in the manner shown in the drawing, to see the position to each other of the contained bloodvessels and the brachial plexus of nerves.

ARTERIES OF THE AXILLA.

The connections of the trunk of the axillary artery, and the distribution of most of its branches, can be studied in Plate ii.

a. Axillary artery.
b. Superior thoracic branch.
c. Acromial thoracic branch.
d. Long thoracic branch.

e. External mammary branch.
f. Subscapular branch.
g. Dorsal branch of the subscapular.

The *axillary artery*, *a*, crosses from the chest to the arm through the axilla; and is limited above by the lower border of the subclavius muscle,

S, and below by the lower edge of the teres major, D. Without dissection, the situation of the vessel may be indicated by a line, on the surface of the body, from a point of the clavicle somewhat on the sternal side of the middle of the bone, to the inner border of the coraco-brachialis muscle, K.

In a dissection carried no farther than the one from which the drawing is taken, the artery is divided into three parts by the pectoralis minor, B, viz., one part above, one beneath, and one beyond the muscle.

The upper or *first part* lies in the axillary sheath between, but deeper than its companion vein and nerves. Superficial to the sheath is the clavicular attachment of the great pectoral muscle; and underneath it is the side of the chest with the intercostal muscles of the first space, and the serratus magnus, F.

Crossing the artery, are some small branches of the companion vein and nerves;—thus directed over it from the outer side is the cephalic vein, *l*, and an anterior thoracic nerve, 3; and passing under it is the nerve to the serratus, 5.

Second part. Here the artery is covered by both pectoral muscles, large and small; but it is without muscular support behind in consequence of its position across the axilla.

The large axillary vein, *h*, has the same relative position to this as to the first part; whilst the brachial plexus, 1, dividing into pieces, is so arranged that one bundle lies outside, another inside, and a third behind the vessel.

The *third part*, twice as long as either of the others, is in contact for two thirds of its length with the pectoralis major, but thence to the ending it is covered only by the common tegumentary structures. It rests successively from above down on the subscapularis, II, the latissimus dorsi, C, and the teres major, D. To its outer side lies the coraco-brachialis muscle, K.

The position of the companion vein remains the same as above; but the connections of the nerves are altered, for the brachial plexus has divided into its terminal branches, which are placed on opposite sides of the vessel. Outside are two nerves, the musculo-cutaneous, 11, reaching only a short distance; and the median, 12, which extends throughout. Inside is the ulnar nerve, 13 (here somewhat displaced); and more or less removed from the artery, is the small internal cutaneous nerve, 9. Superficial to the artery is the large internal cutaneous, 14; and beneath but concealed by it, the circumflex and musculo-spiral nerves—the former reaching only to the edge of the subscapularis muscle.

Number and position of the arterial offsets. Branches are distributed internally to the thorax, and externally to the shoulder and arm.

From the first part come two offsets, the highest thoracic, *b*, and acromial thoracic, *c;* the first is small and irregular in its size and position; and the latter, much larger, springs close to the edge of the pectoralis.

Only occasionally is there any named branch on the second part.

Four or five branches spring from the third part of the parent trunk. The first of these, long thoracic, *d*, is close to the border of the pectoralis minor. The next or subscapular branch arises opposite the lower border of the subscapularis muscle. Two circumflex arteries take origin near the last, but they are concealed by the trunks of the axillary vessels. The last-named branch given off is the small external mammary, *e*.

Ligature of the artery.—The axillary artery may be tied near the clavicle, as well as near the ending (p. 6).

Near the clavicle, or above the small pectoral muscle, the vessel lies deeply, and is reached only after cutting through the pectoralis major. Two offsets, superior and acromial thoracic, spring usually from this part of the artery, with the supra-scapular (a branch of the subclavian) sometimes, and they leave scarcely interval enough for the application of a ligature, especially if the first is large. The connections also of the artery with superficial vessels and nerves are so complicated (see Plate) as to render hazardous ligature of it at this spot.

The vessel might be tied in this situation for aneurism of the lower part of the arterial trunk, or for the arrest of hæmorrhage after an operation high up the arm; but the difficulties in securing the vessel, and the chances of recurring bleeding, may almost deter a surgeon from having recourse to the operation.

Should it be necessary to ligature the artery here, a practical knowledge of the anatomy will assist the operator in his attempts to secure the vessel.

With the arm outstretched, the position of the artery will be marked by a line over the surface of the pectoralis major, which has been described already (p. 20).

The surface depressions on the sides of the clavicular attachment of the pectoralis major being taken as the limit of the incisions, the operator divides by a transverse cut near the clavicle the integuments and the thin platysma muscle, and afterwards the clavicular part of the pectoralis, looking for the cephalic vein at the outer edge of the muscle. When the thick fleshy fibres of the pectoral muscle are cut through, the subjacent fat with small veins, arteries, and nerves, ramifying in it, will appear. With much caution the surgeon finds his way amidst these dangers to the axillary sheath, V, which he opens to the necessary extent.

In the bottom of the wound the firm white brachial plexus of nerves will conduct now to the artery deeply placed between, and overlapped by the nerves and the axillary vein. The artery will be recognized by its pulsation, feel, and color; and when it is detached from the contiguous parts, the operator may enter the aneurism needle between the vein and artery, so that the point of the instrument may be directed towards the nerves as it turns under the arterial trunk.

Aneurism of the upper part of the axillary artery is a formidable disease. It may be confined to the axilla, enlarging forwards and backwards where there is least resistance, or it may pass the bounds of that space, and project above the clavicle into the neck. As long as the disease is low on the vessel, and is confined to the axilla, ligature of the end of the subclavian artery has been resorted to in its treatment. But when it rises above the collar-bone, and the subclavian operation is rendered unsuitable or impracticable, surgeons have sometimes had recourse to the extreme measure of amputating the limb at the shoulder-joint, as there "seemed to be no alternative," to use the words of Professor Syme.

In the last-mentioned class of cases, which are so embarrassing to treat, Professor Syme recommends, that the aneurism should be laid open, and the contents removed, as in the old plan of operating on blood-tumors. From the result of two cases treated successfully in this way, he hopes that "axillary aneurism not amenable to ligature of the subclavian artery may be remedied by the old operation;" and he thinks that, even in cases where ligature of the subclavian is practicable, the plan recommended may be preferable.*

Branches of the artery. All the branches are distributed to the chest and the shoulder, and maintain the circulation in the limb when the

---

* See a Paper, before referred to, on the Treatment of Aneurism, in the Medico-Chirurgical Transactions of London for 1860.

parent vessel has been obliterated. The number of the named branches has been estimated differently by anatomists in consequence of their irregularity.

The *highest thoracic*, *b*, is the smallest branch, and ends on the top of the chest, above the pectoralis minor.

The *acromial thoracic*, *c* (humeral thoracic, thoracic axis?), supplies three sets of offsets, viz., external or acromial, internal or thoracic, and middle or ascending. The outer set enters the deltoid muscle; the inner set is furnished to both pectoral muscles, a few twigs reaching the side of the chest; and the middle set courses over the axillary sheath to the subclavius, and the pectoral and deltoid muscles.

The *long thoracic branch*, *d*, arises opposite the lower border of the small pectoral muscle, and courses along it to the fifth or sixth intercostal space, where it ends in the surrounding parts, and communicates with the intercostal arteries. In the female, it supplies the breast.

The *subscapular*, *f*, a large branch, passes along the muscle of the same name to the inferior angle of the scapula, and is distributed by large branches to the contiguous muscles, serratus and latissimus, anastomosing in the first with the intercostals.

Near its beginning, the dorsal scapular branch, *g*, leaves it to supply the opposite surfaces of the scapula. See Plate v.

Two *circumflex arteries* encircle the humerus, meeting on the outer side. Plate v. may be looked to for a delineation of them.

Other *muscular offsets* (not marked by letters of reference) enter the coraco-brachialis muscle.

Two occasional branches are noticed below, viz., the alar thoracic and external mammary.

Alar thoracic. This belongs to the glands in the axilla, and is seldom to be found as a distinct branch (Quain): offsets to the glands are generally supplied by the subscapular. Plate i. If the alar thoracic exists as a separate artery, it may spring from the second or the third part of the axillary trunk.

The *external mammary*, *e*, appears to be a compensating branch to the long thoracic, *d*, both supplying like parts. It begins near the termination of the axillary trunk, and is accompanied by a vein, *m*.

Anastomosis of the branches. The blood finds it way from one part of the body to another through the communications of the smaller vessels, though its flow in the main trunk is obstructed, and the anastomoses of the branches of the axillary artery with those of the neck and chest, by which the collateral circulation would be established after ligature of the axillary artery, will be now considered.

On the chest, the thoracic offsets of the upper thoracic, acromial and long thoracic, external mammary and subscapular branches anastomose with the intercostal and internal mammary arteries.

On the shoulder, the branches of the axillary communicate with two branches of the subclavian trunk, viz., the posterior scapular and supra-scapular. Offsets of the subscapular artery, distributed to both surfaces of the scapular, join both the above-mentioned subclavian branches. Other anastomoses take place with the supra-scapular in the following way: through the deltoid muscle offsets of the acromial, thoracic, dorsal scapular, and posterior circumflex communicate with that artery, and through the capsule of the shoulder-joint, the anterior and posterior circumflex unite with it.

VEINS IN THE AXILLA.

All the smaller companion veins which would interfere with the view of the arteries and nerves have been taken away.

h. Axillary vein.
k. Brachial vein.

l. Cephalic vein.
m. External mammary.

The *axillary vein*, h, has the same extent as the artery by whose side it lies, and is continuous in the limb with the superficial vein called basilic. Plate iii. Throughout its length it maintains the same position with regard to the artery, *i. e.*, on the thoracic side, and it has similar connections with the parts around. Below the pectoralis minor the vein is often double, and above that muscle it has been once found divided (Morgagni).

Contributing small veins, corresponding with the branches of the artery, enter it at intervals; it receives besides near the lower border of the subscapularis muscle a trunk, k, formed by the brachial veins, and near the clavicle, the superficial vein of the arm—cephalic, l.

The *cephalic vein*, l, ascending over the shoulder between the pectoral and deltoid muscles, sinks through the fascia of the limb, and passing under the great pectoral, pierces the axillary sheath to reach its destination. Its position to the axillary artery has been specially described.

NERVES IN THE AXILLA.

With the exception of one lateral cutaneous nerve of the thorax, all the nerves here represented are derived from the brachial plexus.

1. Brachial plexus.
2. Thoracic offsets of the plexus.
5. Nerve to the serratus magnus.
6. Nerve to the latissimus.
8. Lateral cutaneous of the second intercostal.
9. Small internal cutaneous.
10. Nerve to the teres major.
11. Musculo-cutaneous nerve.
12. Median nerve.
13. Ulnar nerve.
14. Large internal cutaneous of the arm.

The *brachial plexus* of nerves, 1, furnishes offsets to the chest, shoulder, and arm. Placed on the outer side of the first part of the artery, it surrounds the second part with its large trunks, and terminates in branches for the arm, which lie around the third part of that vessel. The following are its offsets to muscles bounding the axilla.

Anterior thoracic nerves are two or three in number. Two, 2, 3, come from the outer part, and one, 4, from the inner part of the plexus, and supply the pectoral muscles; the small pectoral receives its offsets at the under surface from the nerve marked 4.

Nerve to the serratus magnus, 5, comes from the plexus above the clavicle, and may be seen ramifying in its muscle.

Nerve to the latissimus dorsi, 6, enters opposite the axilla the under surface of its muscle.

The *nerve to the teres major* and *subscapularis,* 10, belongs specially to the first muscle, giving only a small piece to the latter, for the subscapularis is supplied higher up in the axilla by an offset of the plexus.

The remaining branches of the brachial plexus are continued to the upper limb, viz.:

The *small internal cutaneous,* 9 (nerve of Wrisberg), communicates in the axilla with the second intercostal nerve, 8.

The *musculo-cutaneous,* 11, pierces the coraco-brachialis. The *median,* 12, the *ulnar,* 13, and the *large internal cutaneous,* 14, lie by the side of the axillary artery, and will be traced afterwards in the limb. The remaining two branches of the plexus, circumflex and musculo-spiral, are concealed by the great axillary vessels.

Remarks on the plexus.—A tumor in the axilla compressing any of the surrounding nerves may occasion pain or dulness of feeling, according to the degree of injury, in the part to which the nerve or nerves affected by it may be distributed.

In dislocation downwards of the humerus into the arm-pit, the head of the bone pressing on the nerves which lie along its inner side, occasions the pain or numbness in the limb.

In the flap amputation of the shoulder-joint, the large vessels and nerves are cut last, as the knife forms the flap on the inner side, and the nerves, not contracting after division like the other structures, reach to the end of the flap, and may be involved in the cicatrix left after the wound is healed if they are not cut shorter.

One *lateral cutaneous nerve* of the thorax, 8, has been left to show its connections with the nerve of Wrisberg, 9. Sending a communicating offset to this nerve in the axilla, it is continued onwards to the integuments of the arm as the intercosto-humeral.

LYMPHATIC GLANDS IN THE AXILLA.

One of the highest of the group of lymphatic glands by the side of the axillary vessels is shown in position on the side of the chest. Two small lymphatic vessels ascend from it, and pierce the inner side of the axillary sheath to join the deep lymphatics of the neck.

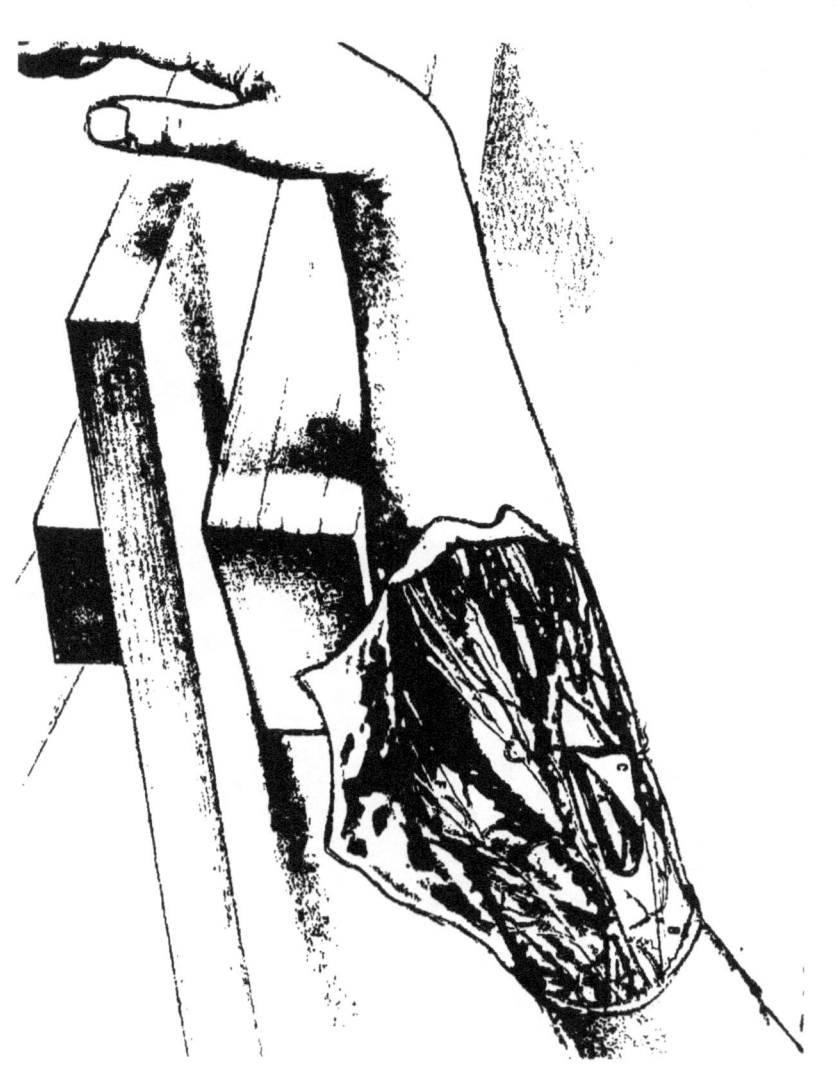

DESCRIPTION OF PLATE III.

A DISSECTION of the superficial veins and nerves in front of the bend of the elbow is represented in this Plate, for the purpose of illustrating the operation of blood-letting.

For the dissection a longitudinal incision was carried over the middle of the joint, and was limited by a transverse cut at each end. On reflecting the two flaps of skin, the subcutaneous vessels and nerves will be found in the fat. A piece of the deep fascia should be raised, as may be seen in the drawing, to show the position of the deep artery and nerve.

BICEPS MUSCLE AND THE FASCIA OF THE ARM.

The deep fascia of the limb deserves special attention, as it is the only protecting layer between the cutaneous veins and the main artery of the arm.

A. Biceps muscle.
B. Deep or special fascia of the arm.
C. Piece of the deep fascia reflected.
D. Inner intermuscular septum.
F. Projection of the inner condyle of the humerus.
I. Intermuscular space on the front of the forearm.

Biceps muscle, A. At its lower end the muscle diminishes in size, and becoming tendinous, is fixed into the radius. Higher in the arm it gives rise to the well-known prominence, with a groove or hollow on each side lodging the superficial veins of the arm, viz., the basilic, h, on the inside, and the cephalic, k, on the outside. The swell of the muscle serves as a guide to the brachial artery along its inner edge.

The *deep fascia*, or the aponeurosis of the limb, invests closely the arm, and is pierced here and there by the nerves and vessels of the integuments. Its component fibres take different directions, some being transverse, others oblique; and it is joined at spots by offsets from the tendons of the muscles. One such offset, added to it from the tendon of the biceps in front of the bend of the elbow, gives it increased strength between the deep artery, *l*, and the superficial median basilic vein, *g*.

On each side of the arm is a thickened part, which is fixed to the humerus between the flexor and extensor muscles, and is called intermuscular septum: these processes are attached to the condyloid ridges of the bone; and the inner one, best developed, is marked by the letter D.

Near the bend of the elbow, where the piece of the fascia is reflected, the contiguity of the underlying brachial artery may be observed.

The fascia is prolonged over the muscles to the forearm; and appearing through it below the elbow is a well-marked yellow line, I, pointing to an intermuscular space which contains the upper end of the radial vessels.

Straightening the elbow-joint increases, and bending the joint relaxes the tightness of the fascia. So the pain consequent on tension of the fascia from accumulation of blood or other fluid beneath it, or from swelling of the parts inclosed by it, may be relieved by placing the limb in a bent position.

SUPERFICIAL VEINS OF THE ELBOW.

Great irregularity prevails in the arrangement of the superficial veins in front of the elbow. The condition of them depicted in the Plate is not quite usual, though it is sufficiently regular for the purpose of describing their anatomy.

a. Median vein of the forearm.
b. Anterior ulnar veins.
c. Posterior ulnar veins.
d. Radial vein of the forearm.

f. Median cephalic vein.
g. Median basilic vein.
h. Basilic vein of the arm.
k. Cephalic vein of the arm.

The *median vein*, *a*, lies along the middle of the forearm, and divides near the bend of the elbow into two, viz., an outer, the median cephalic vein, *f*; and an inner, the median basilic vein, *g*, into which the other veins of the forearm open. At its ending the median communicates with a deep vein through the fascia.

Anterior and *posterior ulnar veins*, *b*, and *c*, gather the blood from the opposite surfaces of the inner half of the forearm, and both join the median basilic, *g*,—the anterior ulnar entering about the middle, and the posterior ulnar at the ending of that vein.

The *radial vein*, *d*, ramifies on the back, and outer part of the forearm, and opens into the end of the median cephalic, *f*. Oftentimes this vein is very small; or it may be wanting.

The *median cephalic vein*, *f*, reaches from the point of splitting of the median, *a*, to the outer border of the limb, where it unites with the radial, *d*, and forms the large cephalic vein, *k*. It crosses the limb obliquely in the hollow between the prominent biceps and the external muscles of the forearm. Underneath it lies the large external cutaneous nerve, 3, and over it pass some offsets of the same nerve. Generally this vein is the smallest of the two pieces into which the median divides, and is sometimes absent.

A moderately tight bandage round the limb just above the elbow, as in the operation of bleeding, does not stop the flow of blood in the median cephalic vein in a muscular arm in consequence of the projection of the biceps arresting the pressure of the band. But the current of blood in the vessel may be commanded by the thumb inserted into the hollow outside the biceps, and pressed downwards steadily.

The *median basilic vein*, *g*, is directed inwards from the median vein, *a*; and uniting with the posterior ulnar veins, *c*, gives rise to the basilic vein, *h*. Usually longer and larger than the median cephalic, it is commonly more transverse in its direction, and is firmly supported by the subjacent fascia and muscle. Joining it below are the anterior ulnar veins. The chief branches of the large internal cutaneous nerve, 1, lie under, and smaller offsets over the vein; but in this dissection the main part of the nerve was superficial to the vein. In the line of the yellow space, I, under the fascia, the brachial artery, *l*, crosses underneath the median basilic vein, the two being separated only by the aponeurosis of the limb somewhat thickened by the prolongation from the tendon of the biceps.

The vein being well supported underneath, the current of blood in it can be readily stopped by the thumb or finger, or by a band round the arm above the elbow compressing the basilic vein.

The *basilic vein*, *h*, begins at the point of union of median basilic, *g*, with the posterior ulnar veins, *c*. Ascending through the lower part of the arm in the groove or depression inside the biceps, it sinks under the fascia half way up the arm, and becomes the axillary vein.

The *cephalic vein* of the arm, *k*, formed, as before said, by the junction of the median cephalic, *f*, with the radial vein, *d*, continues on the outer side of the biceps as far as the shoulder, and ends in the axillary vein. See Plate ii., *l*.

Blood-letting is practised commonly in the veins in front of the elbow. Either the median basilic, *g*, or the median cephalic, *f*, is selected for venesection according to its size; and the median basilic is most frequently opened in consequence of its being the largest, and on account of the readiness with which it may be fixed and compressed against the firm supporting parts beneath. If the operation is to be performed by the student for the first time, the following directions may be of use.

To stop the flow of blood in the superficial veins, a narrow band or fillet is to be tied around the arm from two to three inches above the elbow. This band should not be drawn too tightly, as moderate pressure will arrest the current of blood in the veins; and too great tightness will compress the brachial artery in thin persons, and prevent the free entrance of the blood into the limb below the elbow. After the bandage has been applied, the state of the arteries should be examined, to ascertain that the pulse beats with the same force and frequency as in the other arm; for if the pressure diminishes the current of blood in the main artery, a full stream will not be maintained through the opening made into the vein.

Supposing the median basilic vein, *g*, to be selected for venesection, the position of the brachial artery is to be ascertained by the pulsation, and the vein is not to be opened directly over the beating artery. After this examination the operator stands on the inner side of the limb and grasps the forearm near the elbow with his hand, placing the thumb in front; and, using his left hand for the right arm, and the opposite, he will hold the lancet in the left hand when taking blood from the left limb. With slight pressure of the thumb the vein is now to be fixed; and if this step is omitted, the point of the lancet only punctures and pushes aside the full and freely movable vein. The aperture into this vessel is to be made close to the thumb, both the skin and the vein-wall being divided obliquely to the same extent; and it should be large enough to prevent the blood clotting, and closing it too soon. To give the necessary size (about a quarter of an inch) the lancet is first to be pushed downwards, and is next to be made cut its way to the surface, in order that the structures may be divided from within out; for if the point of the instrument is thrust in and drawn out, making a punctured wound, only a very small quantity of blood will flow through the opening before this is narrowed or stopped by coagulating blood. As the walls of the vein are approximated by the compression of the thumb, too deep an incision of the lancet may cut through the vein, causing effusion of blood beneath with resulting obstruction to the issuing current; and the operation may be accompanied by puncture of the subjacent brachial artery.

The operator does not relinquish his hold of the arm and his control of the vein (for only a few drops of blood will escape till the thumb is removed) until he has had time to put his lancet away, and bring the receiving basin into the proper position. After instructing the person being bled, not to move the arm with the view of trying to direct the jet of flowing blood, he takes his thumb off the vein, and allows the blood to issue in a full stream, though he still supports the limb with his own hand. Leaving the control of the limb to the patient, as when a stick is grasped by the hand, will oftentimes cause the flow of blood to cease; because in his attempts to direct the current of blood into the basin he alters the position of the arm, and the opening in the vein is closed by the skin being brought over it.

Should the displacement of the skin take place, the blood accumulates under it, forming a tumor called "thrombus," and compresses the vein.

When sufficient blood has been obtained the thumb is to be placed on the vein, as before, close below the opening, for the purpose of stopping the bleeding, and the bandage is to be loosened. A small compress of linen, made ready before the operation is begun, is to be placed on the wound; and is to be fixed in position by the fillet applied like a figure of 8 around the elbow whilst the limb is slightly bent. Slight pressure of the bandage, a half bent state of the elbow, and rest, are most conducive to the healing of the wound.

If the median cephalic, *f*, should be selected for venesection in consequence of its greater size, the steps to be taken in the operation are the same as those above referred to, with the exception of the manner in which the current of blood in it is to be checked. Tying up the limb in the usual way will scarcely make pressure enough upon the median cephalic in a muscular arm, because the vein sinks into the hollow on the side of the biceps. A more effectual compression may be exerted by sinking the thumb in the groove between the biceps and supinator longus muscles;

or if a fillet is used, by inserting under it a small compress over the situation of the vein. In consequence of its position in a hollow, the vein may be rather more difficult to reach with the lancet, especially in a fat person.

From the position of the brachial artery under the median basilic vein puncture of it may take place in the operation of bleeding. This serious accident is occasioned by cutting the vein directly over the artery, and pushing the lancet too deeply after transfixing the vein. Injury of another artery may ensue under the following circumstances. One of the large arteries of the forearm (radial or ulnar) may arise higher in the arm than usual, and in passing the elbow to its destination, may lie superficially—being placed generally under the aponeurosis of the limb, but sometimes in the fat, by the side of the veins.* When it is contained in the integuments, its projection in a fat arm might be taken for the swell of a vein on an insufficient examination. The occasional existence of such a state of the arteries should lead to a careful examination of the front of the elbow before venesection, with a view of detecting pulsation not only in the brachial trunk, but also in any other unusually placed artery.

Injury of an artery in blood-letting would be manifested by the blood being redder than ordinary venous blood; by the fluid escaping in jerks; and by pressure on the vein below the opening not stopping the bleeding. Such an untoward accident should be met by placing a conical compress on the wound; and by applying a bandage firmly along the limb with the intention of preventing the escape of the blood, and its accumulation under the deep fascia.

As the wound in the artery does not heal readily, like that in the skin and the vein for instance, a blood-tumor or aneurism usually follows. Into this tumor the blood passes through the hole in the artery, and it is inclosed in a sac formed by the surrounding parts (false aneurism).

Or the wound in the back of the vein not healing, a permanent communication with the artery is established, through which the arterial blood is driven into the vein, producing distention, and a varicose condition of the superficial veins below the elbow. If the edges of the contiguous openings in the vessels unite without the intervention of any sac, so that the vein receives blood directly from the artery, the term aneu-

---

* Surgical Anatomy of the Arteries, by Professor Quain.

rismal varix is applied to that condition of the parts. If, on the contrary, a sac or tumor is formed between the artery and vein, which communicates with both, and serves as a channel by which the arterial current can pass into the vein, the aneurism is called varicose.

For the treatment of a blood tumor or aneurism formed after bleeding, whether it opens only into the artery (traumatic false aneurism) or joins both the artery and the vein (varicose aneurism), an operation on the brachial artery will be needed if its enlargement cannot be controlled by pressure. And the operation suited for the cure of the disease would be that of opening the tumor, and applying a ligature above and below the wound in the artery. If the tumor is somewhat solidified by the deposition of laminated fibrin in it, ligature of the brachial artery in the middle third of the arm would be had recourse to by some surgeons. But the safer practice seems to consist in tying the vessel at the wounded part as a rule; and this treatment would be most suitable also for aneurism connected with a wound of the radial or the ulnar artery in consequence of its unusual origin, and its superficial position in the fat in front of the elbow. Professor Syme advocates cutting down upon the tumor in aneurism from a wound of the brachial in front of the elbow. He says: "I have treated all the aneurisms at the bend of the arm, resulting from wound of the humeral artery through venesection, which have come under my care, amounting to ten in number, by opening the sac, and applying ligatures on both sides of the aperture."*

In the aneurismal varix equable pressure on the limb, which will check the arterial blood entering the tube of the vein to any great extent, may do away with the necessity of any operative proceeding. Should the disease be a source of suffering, and interfere with the use of the arm, as in a laboring man for example, it may be readily cured by ligature of the artery at the part wounded.

In venesection puncture of a nerve will sometimes cause great pain. In the Plate several branches of the internal cutaneous nerve cross the median basilic vein, and any of these might be injured; but as their position cannot be ascertained during life, no precaution can be taken to avoid them. Commonly the puncture occasions only pain at the time of

---

* The Paper on the Treatment of Aneurism before referred to. Medico-Chirurgical Transactions, 1860.

bleeding, though in some conditions of the body it may give origin to serious general disturbance of the health.

Inflammation of the vein or phlebitis may result from bleeding; it will require the treatment appropriate to that affection.

Several other diseased states produced by venesection, with their treatment, were described by Abernethy; and the student who is desirous of obtaining further information may look to the essays of that surgeon.*

The student should observe scrupulously the injunction—never to bleed with a lancet that has been used for other purposes.

BRACHIAL ARTERY AT THE ELBOW.

The lower end of the brachial artery, *l*, which lies under the superficial veins, and may be wounded in venesection, has been laid bare by reflecting a piece, C, of the deep fascia.

In this situation the artery is very near the surface of the limb, and is covered only by the integuments and the deep fascia, B. Along its outer side is the biceps muscle, A, which will serve as the guide to the vessel. Underneath it lies the brachialis anticus muscle (Plate iv., F).

One large accompanying nerve, median, 8, is placed on the inner side of the artery, and the median basilic vein crosses over it.

Only superficial offsets are furnished to the integuments from this part of the vessel.

Ligature of the artery at the elbow may be necessary in consequence of a wound with a lancet in venesection, or with any other cutting instrument.

In the case of a wound from accident the vessel requires to be secured by one thread above and another below the injury; and with the surrounding textures infiltrated with blood, the surgeon may experience some difficulty in finding the ends of the vessel, unless he has studied the connections, and practised previously the operation of applying a ligature to the artery in the dead body.

In an operation here for aneurism after a wound, as when the vessel is punctured in venesection, the tumor is to be opened, and the contents of the sac being removed, the arterial trunk is to be tied above and below the opening in it.

---

*Surgical Observations on Injuries of the Head and on Miscellaneous Subjects, by John Abernethy, F.R.S.; 4th Edit., p. 135: London, 1825.

Cutting down to the artery in front of the elbow is an easy operation in the dead body. Taking the inner edge of the biceps muscle as the superficial guide to the position of the vessel, an incision two or three inches in length, and paralled to the artery, may be carried along the biceps, so as to divide the integuments; and should the median basilic vein come into view at this stage, it may be drawn inwards. The deep fascia is next to be cut to the same extent, and the wound is to be moved inwards over the line of the artery.

Deep in the wound the firm white median nerve appears on the inner side of the artery, but gradually inclining away from it in front of the elbow-joint; this nerve will serve as the deep guide to the position of the vessel, though the operator should be aware that it may be placed away from the artery, lying along the inner intermuscular septum of the arm.* The nerve being recognized, the artery is to be sought between it and the edge of the biceps.

Lastly, the sheath of the vessel having been opened, and the venæ comites separated from the artery, the aneurism needle may be passed, and the ligature may be tied in the usual way.

Some unusual conditions of the arteries in front of the elbow deserve consideration with reference to the operation of blood-letting. The occasional presence of an artery in the fat with the superficial veins has been before noticed, p. 34. The number of large arteries too beneath the fascia may vary. Commonly there is only one, the brachial; but there may be two, which consist of the brachial trunk and the radial or ulnar; and lastly, three may be occasionally found, resulting from division of the brachial into its usual arteries rather above the elbow-joint, and the unusual origin of the interosseous from the brachial high in the arm.† The possibility of so many arteries being present in one spot must suggest caution to the student about to bleed, and to the surgeon undertaking the operation of placing a ligature on a wounded artery in front of the elbow.

There is another unusual state of the brachial artery which would give rise to unlooked-for hæmorrhage from a wound in the lower half of the arm. For instance, the artery leaves sometimes the edge of the biceps,

---

* I have met with three examples of this condition in the dissecting-room of University College. In another body the nerve was deeper than the artery, and was covered, above the elbow, by fibres of the brachialis anticus.

† The facts here referred to shortly, are stated fully in the Surgical Anatomy of the Arteries by Professor Quain, p. 259.

and courses, with or without the median nerve, along the line of the inner intermuscular septum, D. At the elbow it returns to the middle of the limb through the origin of a wide pronator teres muscle, or round a projecting bony point of the humerus (Quain). In such a deviation in the course of the artery, a wound near the elbow on the inner side of the arm, far removed from the line of the biceps muscle, might open this large trunk, and give origin to most alarming, if not dangerous hæmorrhage.

NERVES BEFORE THE BEND OF THE ELBOW.

The anterior cutaneous nerves of the forearm cross the superficial veins in front the elbow in coursing to their destination.

1. Large internal cutaneous nerve.
2. Small internal cutaneous, or the nerve of Wrisberg.
3. External cutaneous nerve.
4. Anterior part of the large internal cutaneous.
5. Cutaneous offsets to the arm of the internal cutaneous.
6. Posterior part of the internal cutaneous.
8. Median nerve.

The *large internal cutaneous nerve*, 1, enters the fat about midway along the arm, and divides into two parts:—One, 4 (the anterior part), is continued along the front of the forearm to the wrist; the other, 6 (posterior part), ramifies on the back of the forearm on the ulnar side, reaching to the lower third. The primary branches of the nerve lie generally under the median basilic vein, instead of over it as in this dissection.

Near the arm-pit a cutaneous offset, 5, leaves the trunk to supply the integuments over the biceps muscle.

Small internal cutaneous nerve, 2, or the nerve of Wrisberg. The origin of the nerve is seen in the arm-pit in Plates i. and ii. It pierces the fascia internal to, and lower down than the large cutaneous nerve, 1; and it ends in the integuments over the back of the elbow. Offsets are directed backwards to the fat and skin of the lower part of the posterior surface of the arm; and one or two communicate with the large internal cutaneous nerve.

In this body the nerve was large, and was placed rather farther forwards than usual.

The *external cutaneous nerve*, 3, or the terminal part of the musculocutaneous (Plate ii., 11), appears at the bend of the elbow beneath the

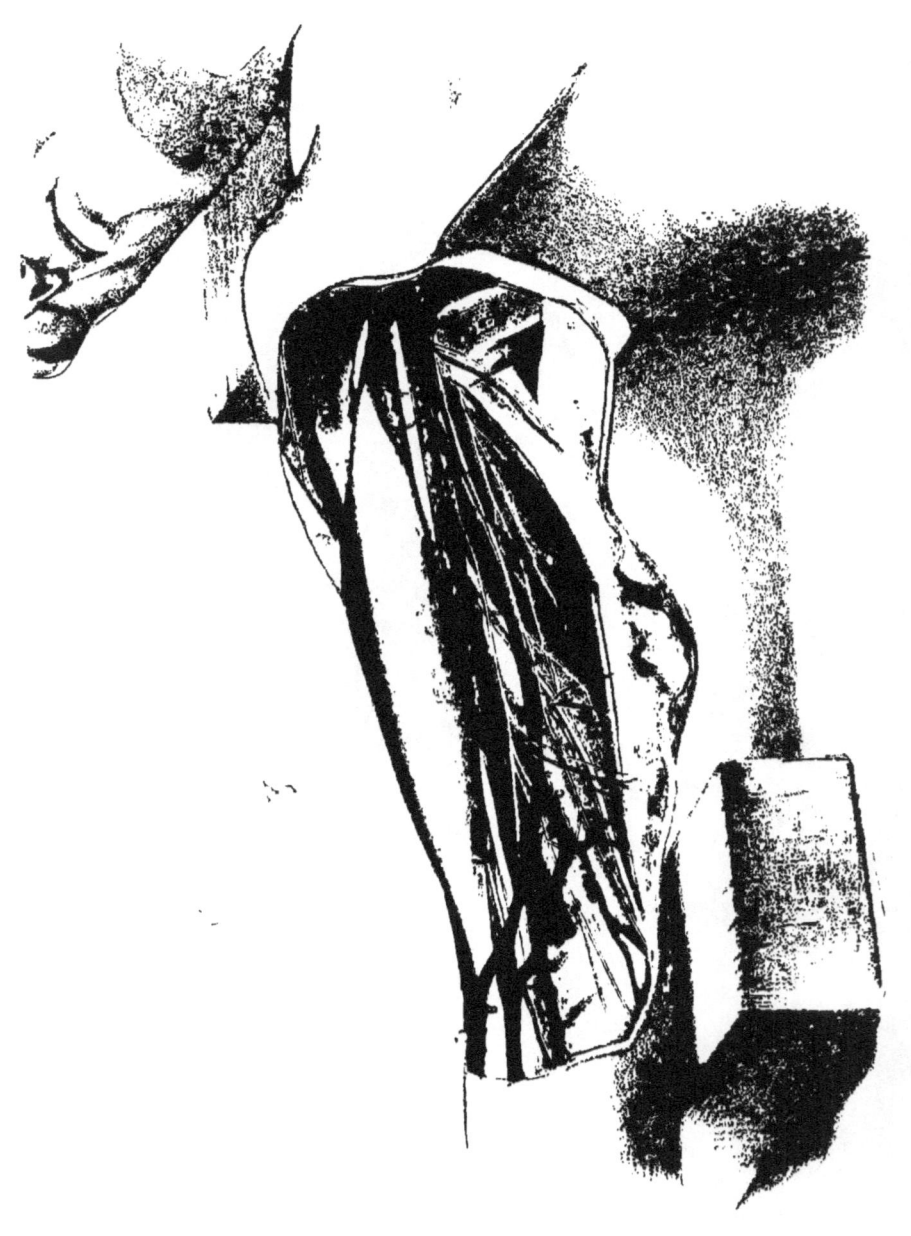

median cephalic vein, f, and its distributed along the radial side of the forearm as far as the ball of the thumb.

The *median nerve*, 8, is continued to the fingers. At the upper part of the dissection it lies inside and near the brachial artery, but opposite the bend of the elbow it begins to incline inwards from that vessel. In the lower as in the upper part of the arm the nerve serves to guide the surgeon to the situation of the large vessel of the limb.

DESCRIPTION OF PLATE IV.

THE relative position of the muscles, vessels, and nerves of the inner side of the arm, after the removal of the integuments and the deep fascia, is shown in this Plate.

The skin may be reflected in two flaps to the sides, by an incision along the centre of the arm, with a cross-cut at each end. In the fat the superficial nerves and vessels are to be found; and then the remains of the fat, and the deep fascia, should be taken away. A small part of the fascia has been left near the elbow, for the purpose of marking its position to the superficial veins and the lymphatic glands.

SURFACE MARKING OF THE ARM.

Along the front of the arm is the well-marked muscular prominence so evident in the Plate. Before the removal of the integuments and fascia, this eminence seems to the feel to be formed by one muscle; but after the dissection has been made, it will be seen to consist of the biceps, D, and coraco-brachialis, H, which may be traced upwards under the anterior fold, R, of the arm-pit. As the chief muscle, D, acts as a flexor of the elbow-joint, it becomes much enlarged in persons occupied, like blacksmiths, in bending the elbow.

On each side of the swell of the muscles is a surface depression: the two meet below in a hollow in front of the elbow, which contains the superficial veins and nerves; but above they separate, the inner one joining the arm-pit, and the outer one subsiding at the insertion of the deltoid muscle, S.

In the inner depression, which is most marked, lies the basilic vein, *g*, with the large internal cutaneous nerve, 4, and lympathics: these are contained in the fat, and are usually distant a short way from the edge of the biceps. Beneath the fascia of the limb and close to the muscle are lodged the brachial vessels, *k*, and the companion median nerve, 7.

The outer depression is less wide and deep, and corresponding with it is the superficial vein, the cephalic, *h*. In it, towards the elbow, the external cutaneous nerve of the forearm makes its appearance though the fascia (Plate iii. 3).

Wounds in the outer bicipital hollow may be large and deep without injuring any important part; whilst in the inner one scarcely a puncture can be made without endangering some vessel or nerve. The issue, seton, and cautery are applied usually at the top of the outer bicipital groove, just below the insertion of the deltoid muscle, because the spot is free from any active subjacent muscle to give rise by its contractions to pain in the sore that has been produced.

MUSCLES AND FASCIA OF THE ARM.

The muscles on the front of the humerus which pass over the elbow, viz., the biceps, D, and the brachialis anticus, F, bend the elbow-joint by bringing forwards the bones of the forearm, to which they are fixed. Behind the humerus is a large three-headed or tricipital muscle which is attached to the ulna, and, drawing backwards that bone, acts as an antagonist to the flexor muscles.

A. Fascia of the forearm.
B. Offset to the fascia from the tendon of the biceps.
C. Inner intermuscular septum of the arm.
D. Biceps flexor brachii muscle.
F. Brachialis anticus muscle.
H. Coraco-brachialis muscle.

K. Inner head of the triceps extensor muscle.
M. Middle head of the triceps.
N. Teres major muscle.
P. Latissimus dorsi muscle.
R. Pectoralis major muscle.
S. Deltoid muscle.

The *deep fascia* of the arm is continuous with that of the forearm, A, and is attached to the prominences around the elbow. C marks the inner intermuscular septum of the arm, which is inserted into the condyloid ridge of the humerus, and gives origin in front to the brachialis anticus, F, and behind to the inner head of the triceps, K. In front of the septum a piece of the fascia has been left with superficial lymphatic glands on it. At B, an offset from the tendon of the biceps joins the fascia.

Biceps brachii muscle, D. The origin by two heads from the scapula, is shown in Plate ii. Half way along the arm the heads blend in a fleshy belly; and the muscle is inserted below by a tendon into the tubercle of the radius, after giving a fibrous process, B, to the deep fascia. A third slip or head arises occasionally from the middle of the humerus; and if it crosses over the main vessels, as it is directed outwards, it may complicate the operation of tying the artery.

Except at the origin and insertion the muscle is superficial; and it covers partly the other two muscles in front of the humerus, viz., the coraco-brachialis, H, and brachialis anticus, F. Along the inner edge lie the brachial artery, k, and the accompanying veins and nerves; and along the outer edge is the cephalic vein, h.

The muscle flexes the elbow-joint by acting on either the radius or the humerus, according as the one or the other may be free to be moved. It is also a supinator of the hand. And if the radius is fixed it can assist in carrying the limb forwards from the side. As the muscle contracts in the living body the swell of its belly rises towards the pectoralis major.

The *brachialis anticus*, F, arises from the front of the humerus for the lower half of the bone; and from the intermuscular septum on each side, viz., from all the inner one, but from only the upper part of the outer one, some muscles of the forearm excluding it below (Plate xii.). It is inserted into the fore part of the coronoid process of the ulna.

Resting on the humerus and the elbow-joint, it is concealed by the biceps, and vessels and nerves. Sometimes a fleshy slip from it covers the brachial artery or the median nerve at the lower part of the arm.

This muscle reaches over the elbow, and is the chief agent in bending that joint.

The *coraco-brachialis muscle*, H, is shown better in Plate ii., to which reference may be made.

The *triceps extensor cubiti* consists above of three parts or heads; and its anatomy will be given more fully in the description of Plate vi.

The middle head, M, arises from the scapula; and the inner and outer heads are attached to the humerus and the intermuscular septa. The insertion of the muscle into the olecranon process of the ulna will be afterwards seen.

In this view of the parts the middle head lies beneath the teres major, N, and latissimus dorsi, P, and touches the brachial vessels and their companion nerves for one to two inches. And the inner head surrounds the ulnar nerve, 8, and the inferior profunda artery, n: this is more evident in Plate vi.

The *teres major*, N, and *latissimus dorsi*, P, coming forwards to their insertion into the humerus, bound behind the hollow of the axilla (Plate i.).

The *pectoralis major*, R, curves over the muscles of the front of the arm as it passes from the thorax to its insertion into the humerus. At its attachment to the bone, it joins the deltoid muscle, S.

VEINS OF THE ARM.

The superficial veins of the limb diminish in number from the hand upwards. At the elbow they blend into two, which have a constant course on the sides of the biceps muscle to the axilla. A somewhat different arrangement from that in Plate iii. is here noticeable.

- *a.* Median vein of the forearm.
- *b b.* Anterior ulnar veins.
- *c.* Posterior ulnar vein.
- *d.* Median cephalic vein.
- *f.* Median basilic vein.
- *g.* Basilic vein of the arm.
- *h.* Cephalic vein of the arm.
- *s.* Inner companion vein of the brachial artery.

The *median vein*, a, splits in the usual way into two branches, which are directed outwards and inwards to receive the radial and ulnar veins. In this body the anterior ulnar veins, b, b, are large, and join the median basilic, f, at separate points, after being united by a cross branch.

The *basilic vein*, g, formed by the union of the median basilic and anterior ulnar veins near the elbow, ascends in the fat to the middle of the arm; then piercing the deep fascia, it is directed onwards to the axilla by the side of the brachial artery, and becomes the axillary vein at the lower border of the teres major muscle. Soon after it sinks through the fascia it communicates usually with one of the companion veins, s, of the brachial artery.

Cephalic veins, h.—Only the upper part of this vein is visible as it crosses between the muscles great pectoral, R, and deltoid, S, to end in the axillary vein. Springing below from the junction of the median cephalic, *d*, with the radial vein, it ascends in the fat to the shoulder outside the biceps muscle. An unusual superficial artery accompanied it is this dissection.

Venæ comites.—The companion veins of the brachial artery, two in number, lie one on each side of that vessel, and join at intervals by cross branches; the inner one is marked *s* in the Plate. Receiving small veins which accompany the branches of the artery, they join commonly into one at the lower part of the axilla; and this ends in the axillary vein near the lower border of the subscapularis muscle (Plate ii. *k*).

ARTERIES OF THE ARM.

The brachial artery and the end of the axillary trunk may be studied in this dissection with their connections undisturbed. The ramifications or ending of the branches must be learned with the aid of the other Plates.

k. Brachial artery.
* Spot best suited for ligature of the vessel.
l. External mammary branch of the axillary artery.

m. Muscular offset of the superior profunda branch.
n. Inferior profunda branch.
p. Anastomotic branch.

The *brachial artery*, *k*, extends from the lower border of the teres major muscle, N, to a finger's breadth below the bend of the elbow (Quain), where it bifurcates into the radial and ulnar arteries. The inner edge of the muscular prominence of the coraco-brachialis and biceps marks its position in the limb; or a line from the arm-pit to the middle of the bend of the elbow would correspond with the course of the vessel.

In consequence of its superficial position in the arm the vessel can be readily compressed. Above the spot marked with an asterisk the artery lies inside the humerus, and pressure to act on it should be directed outwards against the bone; but below that spot it inclines in front of the bone, and the blood will be stopped in it by forcing backwards the finger or the thumb.

Its connection with muscles and fascia are the following:—Beneath it, from above down, are the long head of the triceps, M; the inner head, K, of the same muscle; the coraco-brachialis, H, where the asterisk is placed, and thence to the ending, the brachialis anticus, F. Superficial to the artery is the deep fascia of the limb with the integuments.

Two companion veins are close to the brachial trunk—one on each side—and anastomose across it after the manner of such veins; and at the bend of the elbow the median basilic vein, *f*, crosses the artery. The basilic vein, *g*, lies inside the line of the vessel—sometimes nearer, and at others farther from it.

Several nerves accompany the artery above, but only one below. The median nerve, 7, keeps close to the vessel throughout, except in front of the elbow, where it inclines away to the inner side; as low as the part marked thus * it is outside the vessel, then it crosses gradually over, though occasionally under the artery, and gains the inner side about two inches above the elbow. The ulnar nerve, 8, lies inside and in close contact with the artery nearly to the asterisk, but at that spot it diverges from the vessel and courses along the inner intermuscular septum. The musculo-spiral nerve is placed behind the upper part of the artery for two inches (see Plate vi.). The large internal cutaneous nerve, 4, rests on the upper third of the brachial artery; but in some bodies it is moved farther in, as in the dissection from which the drawing was taken.

Position and names of the branches. Besides small muscular and cutaneous offsets, four named branches spring from the brachial trunk. The highest and largest, *upper profunda*, leaves the back of the artery about an inch from the beginning. The next largest, the *inferior profunda*, *n*, arises near the upper end of the inner intermuscular septum. A *nutritive artery* of the shaft of the humerus begins near the last, and is covered by the biceps. Another small branch, the *anastomotic artery*, *p*, comes from the parent trunk near the elbow.

All the branches are small except the superior profunda; and no two arise at opposite sides of the trunk to interfere by a cross current with the healing process after a thread has been put on it. Almost any point would therefore be available for the application of a ligature; but the spot generally selected is marked with an asterisk in the Plate, the vessel being here free from any large offset, and being firmly supported by the coraco-brachialis and the humerus.

Ligature at the middle of the artery. This operation on the brachial trunk, without a wound at the spot where it is tied, is sometimes rendered necessary by an aneurism, or by hæmorrhage from a vessel lower in the limb.

Under ordinary circumstances the operation is not difficult, as the brachial trunk is so near the surface, and the guides to the vessel are good. The superficial guide to the position of the artery is the inner edge of the biceps muscle, and the deep guide during the operation is the large median nerve.

When the vessel is to be secured the operator stands on the inner side of the limb, and fixing his eye on the spot thus marked *, makes a cut two or three inches long on the biceps muscle near the inner edge, but not over the vessel. The skin, fat, and deep fascia are to be divided down to the fleshy fibres; and the incision is then to be moved inwards over the line of the brachial artery, the loose skin readily allowing this shifting of its position.

Bending now the elbow, to relax the biceps muscle and allow of its being kept out of the way, the firm median nerve is to be looked for close to the edge of the biceps, where it lies outside the vessel, or is coming inwards over the arterial trunk. The median nerve being found, and the knife having been carried along it to divide its sheath, is next to be drawn inwards from the edge of the biceps with a narrow retractor, but special care must be taken not to draw the artery out of place with the nerve. Within the space limited by the nerve on the one side and the muscle on the other, the operator seeks the artery by cutting away the fat bit by bit.†

Supposing the artery recognized, its sheath is to be seized with the forceps, and a piece is to be cut out, care being taken that the point of the scapel does not injure the vessel beneath. Without loosing the sheath from the forceps a blunt instrument, like the point of a director, may be inserted into the hole of the sheath to separate the artery; and on its withdrawal the aneurism needle is to be carried round the vessel in the same channel. The surgeon avoids detaching the artery from its sheath more than is required for the passage of the needle; for separation of the

---

† Some experience in superintending the operations of students on the dead body has convinced me of the expediency of directing the nerve to be drawn inwards. If this mode of proceeding is not adopted, the beginner comes upon the ulnar nerve and the basilic vein, which he may mistake for the median nerve and the brachial artery.

two destroys the vasa vasorum, occasioning the death of the arterial coats, and, as a consequence, hæmorrhage may follow the coming away of the ligature.

Let the ligature be put on the vessel as high as the sheath is detached; and before tying it, pressure should be used for the purpose of ascertaining whether the circulation through the chief vessels of the limb can be arrested. Should the pulse still beat as before at the wrist, the existence of more than one arterial trunk may be suspected; and the operator, after tying the one, seeks another by its side. If two arteries are present both are to be secured; for the object in view when putting a ligature on the brachial trunk, is to stop the entrance of the blood into the limb through the main vessel, and to insure its coming in only slowly, and through the anastomosing channels.

Before an attempt is made to place a ligature on the brachial trunk, the difficulties likely to arise from different states of the artery or of the surrounding parts should be well considered.

An unusual position of the brachial artery has been observed. In the condition referred to the vessel separates from the biceps above, or about midway between the arm-pit and elbow, and courses through the arm along the inner intermuscular septum, C (p. 35). So, in an operation at the usual spot, if the main blood-vessel cannot be found by the side of the muscle, it should be sought further in, or nearer the inner border of the limb.*

There may be more than one large artery in the limb as before said. Two vessels have been found as frequently as 1 in 5, and the surgeon may expect therefore to meet with this condition.† When two vessels are present they usually lie side by side in the place of the brachial; and their existence might be inferred in an operation in consequence of the smaller size and more superficial position of the vessel first found. But sometimes the two are not in contact with each other: thus, one, the smallest (radial) may lie in the place of the brachial trunk; and the

---

* Two instances of this kind were met with during operations on the dead body, and have been put on record by Mr. Quain : " Commentaries on the Arteries," p. 259. I have observed a similar unusual place of the artery, with difficulty in finding the vessel, whilst I was engaged in superintending the operations of students.

† The Anatomy of the Arteries, by Mr. Quain.

other, the larger artery, may be moved inwards from the edge of the biceps to the inner intermuscular septum.

The depth of the artery varies with different states of the biceps muscle. Sometimes the brachial trunk is covered, at the spot where ligature is practised, by a fleshy slip of origin of the biceps from the humerus. The presence of fleshy fibres over the artery would cause some embarrassment to an operator unacquainted with that fact; and the knowledge of the occasional existence of this condition teaches, that a previous examination of the arm should be made, with the view of detecting it by the difference in the force of the pulsations of the artery.

Change in the position of the median nerve with respect to the brachial artery may bring danger in an operation, as the nerve serves as the deep guide to the vessel. In the ordinary arrangement the nerve is superficial to the artery, and is met with first; but not unfrequently it crosses under the artery, and would not be found so soon as the vessel. When this last-named position of the nerve exists, the danger of wounding the artery or its companion veins is increased in consequence of these being nearer the surface, and being reached unexpectedly.

Branches of the artery.—The offsets of the artery are small and numerous, but only a few have received names. After supplying the muscles and contiguous parts the chief branches course to the elbow, and join offsets of the trunks in the forearm.

The *superior profunda branch* arises from the trunk of the artery above the letter, k, and winds to the back of the arm, where it ramifies in the triceps, and ends at the elbow. (See Plate vii.) One offset is marked, m, in the Drawing.

The *inferior profunda*, n, arises near the spot which is commonly chosen for ligature of the trunk, and runs along the ulnar nerve to the elbow: it anastomoses with the posterior recurrent branch of the ulnar artery.

The *nutritive artery of the shaft of the humerus* arises between k, and *, and entering an osseous canal, supplies the medullary structure of the bone.

The *anastomotic branch*, p, is directed inwards through the inner intermuscular septum, and communicates with the inferior profunda, n. An offset descends in front of the elbow joint, supplying the brachialis anticus and one or more muscles of the forearm, and anastomoses with an anterior recurrent branch from the ulnar artery.

Muscular offsets spring from the trunk at intervals, and supply the muscles on the fore part and the back of the humerus.

Small *cutaneous* offsets to the arm are shown coming from the end of the brachial, and the end of the axillary artery.

Anastomoses of the branches.—After ligature of the brachial artery the blood is conveyed into the limb by the anastomosis of the branches arising above, with those beyond the spot tied. Thus the superior profunda joins behind the elbow with the anastomotic and the recurrent interosseous; and on the outer side with the recurrent branch from the radial artery (Plate vii.). The inferior profunda communicates with the anastomotic, and with the posterior recurrent of the ulnar (Plate viii.). And the anastomotic branch, joining the profunda, transmits its blood to the anterior recurrent branch of the ulnar (Plate viii.). The artery entering the shaft of the humerus will anastomose above and below with the vessels supplied to the ends of the bone.

NERVES OF THE ARM.

All the nerves included in this dissection are derived from the brachial plexus in the axilla, with the exception of the small offsets over the shoulder, which come from the cervical plexus in the neck. Only a part of each trunk is laid bare, as it passes onwards to its destination in the forearm.

1. Internal cutaneous branch of the musculo-spiral.
2. Branch of the musculo-spiral to the inner and middle heads of the triceps.
3. Nerve of Wrisberg or small internal cutaneous.
4. Internal cutaneous (large).
5. Anterior branch of the internal cutaneous.
6. Posterior branch of the internal cutaneous.
7. Median nerve.
8. Ulnar nerve.
9. Branches of the cervical plexus.

The trunk of the *musculo-spiral nerve*, lying beneath the brachial artery, furnishes a cutaneous branch, 1, to the integuments of the back of the arm; this reaches as far as the lower third, or sometimes nearly to the elbow. A muscular branch, 2, to the inner head, K, and the middle head, M, of the triceps, arises in common with the preceding.

The *nerve of Wrisberg*, 3, and the *large internal cutaneous*, 4, pierce the fascia of the arm rather below the middle, and are distributed to the integuments of the back of the arm and forearm: their position internal to the brachial artery may be noticed. Usually the cutaneous nerve, 4, lies over the upper part of the artery. Its place at the elbow under the median basilic vein is regular: for another arrangement, see Plate iii.

The *median nerve*, 7, takes the same course in the arm as the brachial artery, and lies close to that vessel (p. 46). Outside the artery above, and inside below, it crosses over the bloodvessel so as to be found on the inner side about two inches above the elbow. Sometimes the nerve passes under instead of over the artery in its change of position from the one side to the other. No branch is distributed from it in the upper arm.

Being the companion nerve to the main artery, it changes generally its place when the vessel deviates from the usual site. Thus in those instances in which the brachial artery courses along the inner intermuscular septum to the elbow the nerve usually accompanies it; but the nerve may be near the septum without the bloodvessel (p. 35). In this last case a wound just above the elbow might cut through the nerve, and interfere with the actions of the parts supplied by it; or from the close contiguity of the ulnar and median nerves, one being before and the other behind the intermuscular septum, C, the same wound dividing both trunks would cause loss of power in the muscles on the front of the limb below the elbow, with insensibility in the fingers and the palm of the hand, and in the part in the back of the hand.

The *ulnar nerve* passes through the upper arm without branching, and enters the forearm behind the elbow-joint. As far as the middle of the arm the nerve is close to, and rather behind the brachial artery; but it separates afterwards from the vessel, passing through the intermuscular septum, and is continued behind this piece of fascia to the hollow between the olecranon and the inner condyle of the humerus. Pressure applied to it behind the elbow-joint causes a peculiar tingling along the inner side of the hand, and in the inner two fingers.

LYMPHATICS OF THE ARM.

Superficial lymphatics accompany the superficial veins in the arm; and the greater number lie along the inner part of the limb. Above the elbow are some superficial lymphatic glands in front of the intermuscular septum, which are marked thus, † † †; these are the lowest superficial glands in the limb. Three glands were present in the dissection. Enlargement of those glands may be brought on by causes which induce inflammation and swelling of lymphatic glands elsewhere; and a small tumor in this part of the arm may be owing to an increase of one of the glands.

Deep lymphatics with their appertaining glands course with the trunks of the bloodvessels beneath the fascia, and enter the glands in the axilla.

DESCRIPTION OF PLATE V.

This view exhibites the dissection of the shoulder, and that of the superficial muscles and vessels of the back of the scapula.

On the detached limb this dissection will follow the examination of the subscapularis muscle on the under surface of the scapula; and it is readily made by reflecting the integuments and the deep fascia from before backwards towards the lower angle of the blade-bone. By cutting through the deltoid near its upper attachment, the vessels and nerve beneath it can be traced out.

MUSCLES OF THE SCAPULA, SHOULDER, AND ARM.

Three groups of muscles are laid bare more or less completely in the dissection, viz., the muscles of two borders of the scapula; those of the posterior surface of that bone; and those of the shoulder and the back of the arm.

All the muscles passing between the humerus and the scapula are relaxed, and are consequently wide and hanging; but in Plate vi. the muscles are shown on the stretch, where the difference in their form and position may be noted.

The dorsal muscles of the scapula cover the shoulder-joint, and will ceive injury in dislocation of the head of the humerus.

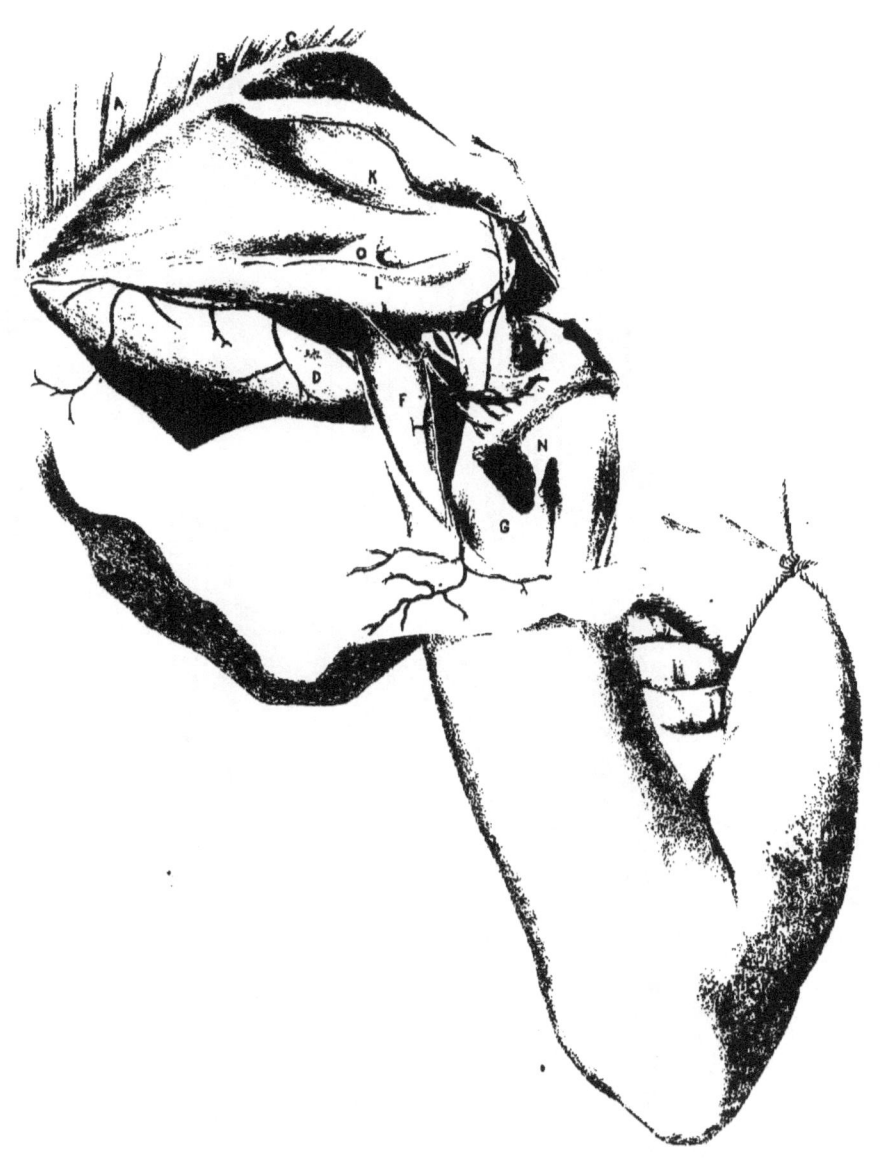

A. Rhomboideus major.
B. Rhomboideus minor.
C. Levator anguli scapulæ.
D. Teres major.
E. Latissimus dorsi.
F. Long head of the triceps.
G. Outer head of the triceps.

H. Supra-spinatus.
K. Infra-spinatus.
L. Teres minor.
N. Deltoid muscle.
O. Fascia on the dorsal scapular muscles.

The three muscles marked A, B, C, arise from the spinal column, and are fixed into the base of the scapula.

The *rhomboideus major*, A, is inserted between the spine and the lower angle of the bone.

The *rhomboideus minor*, B, is attached opposite the smooth surface at the root of the spine.

The *levator anguli scapulæ*, C, is fixed above the last, reaching from it to the upper angle of the shoulder-blade.

From the direction of their fibres the muscles, when acting without the trapezius, would lower the point of the shoulder, by raising and bringing towards the spinal column the lower angle and base of the scapula.

Connected with the inferior border of the scapula are the teres major and the long head of the triceps; and the latissimus dorsi crosses the others, resting on the inferior angle of the bone.

The *teres major*, D, arises from a special impression on the lower angle of the scapula, from the deep fascia covering the dorsal scapular muscles, and from the lower edge of the scapula as far forwards as an inch from the long head of the triceps. It bounds the axilla behind, and lies in front of the long head of the triceps (Plates i. and ii.).

The muscle diverges in front from the axillary border of the scapula, leaving a triangular interval between it and the bone; and it is concealed partly by the latissimus dorsi, E, when viewed from behind.

The *latissimus dorsi*, E, is attached to the lower part of the trunk of the body by the one end, and to the humerus by the other. Winding over the lower angle of the scapula and the teres major, it ascends in front of the teres to its insertion into the bicipital groove (Plate ii.).

In the dissection the muscle slipped down somewhat in consequence of its relaxed condition, but its natural place on the angle of the scapula is displayed in Plate vi.

These two muscles could draw the arm to the scapula if the member was at a distance from the trunk; or if the limb was fixed, as in climbing, they would help to approximate the trunk to the raised limb.

And when the latissimus has drawn the humerus backwards, it will rotate inwards that bone. If the lower end of the raised humerus is not free to move, this muscle acting with the teres and pectoralis major draws down the upper end, and may dislodge the head from the articular surface of the scapula.

The dorsal scapular muscles, H, K, and L, cover the shoulder-joint above and behind, and converge to the head of the humerus. A deep fascia covers the muscles, and gives origin to the fleshy fibres: one piece dips between the two infra-spinous muscles, K and L, and is fixed to the scapula.

The *supra-spinatus muscle*, H, fills the hollow above the spine of the scapula. Arising from the bone and the fascia, it passes over the shoulder-joint to be inserted into the upper impression on the great tuberosity of the humerus.

The *infra-spinatus muscle*, K, is named from its position below the spine of the scapula. It arises, like the preceding, from the underlying bone and the fascia stretched over it; and, crossing the shoulder-joint, it is inserted into the middle impression on the great tuberosity of the humerus.

The superficial fibres from the spine of the scapula and the fascia are directed forwards over the fibres coming from the blade part of the bone.

The *teres minor*, L, arises by the side of the infra-spinatus from the fascia, and from a special impression along the axillary border of the scapula. Covering the joint, it is inserted into the lowest mark on the great tuberosity of the head of the humerus, and into the bone below by a few fleshy fibres.

The three muscles above noticed are called "articular" from touching the joint. When in action they cause the humerus to move in the following directions. If the bone is hanging the supra-spinatus will assist the deltoid in raising the arm; and the infra-spinatus and teres minor acting together will draw backwards the point of the bone to which they are fixed, becoming external rotators. If the humerus is elevated the two last muscles below the scapular spine will assist the deltoid in carrying backwards the arm almost horizontally.

They suffer more or less injury in dislocations of the shoulder-joint.

Should the humerus be dragged downwards from its socket all three may be torn across; or, the muscles remaining whole, a shell of bone, into which they are inserted, may be detached from the humerus. In dislocation backwards the head of the humerus lies under the infra-spinatus, K, and teres minor, L, which are relaxed; and the supra-spinatus is directed backwards, and made tense round the spine of the scapula. But supposing the bone dislocated forwards (on to the other side of the scapula), the infra-spinal muscles will be much stretched if not torn.

The two arm muscles are the deltoid, forming the prominence of the shoulder, and the triceps, which lies behind the arm bone.

The *deltoid muscle*, N, arises from the scapular arch opposite the attachment of the trapezius, viz., from the outer third of the clavicle, and from the acromion and the lower edge of the spine of the scapula as far back as the posterior smooth triangular surface, where it blends with the deep fascia covering the infra-spinous muscles. It narrows below, and is inserted into an impression on the outside of the humerus above the middle. Sufficient of the muscle has been divided to show beneath it the head of the humerus, the insertion of the dorsal scapular muscles, and the posterior circumflex artery, *a*, and nerve, 1.

Between the acromion process and the deltoid muscle, on the one side, and the head of the humerus with the dorsal scapular muscles on the other, is a bursa—one of the largest in the body—which lubricates those surfaces in the movements of the arm. In chronic rheumatic arthritis, when the surrounding capsule and muscles are destroyed, this bursa communicates with the articulation—the deltoid and acromion becoming incasing structures of the shoulder-joint.

When taking its fixed point above, this muscle is the chief elevator of the humerus, and it can carry backwards and forwards the raised limb; but in all these movements it is assisted by the scapular muscles. The arm is raised by the central fibres of the deltoid and the supra-spinatus muscle, H; it is moved forwards by the clavicular fibres and the subscapularis; and it is carried back by the fibres attached to the spine of the scapula, and by the infra-spinatus, K, and teres minor, L.

Supposing the deltoid to act from the humerus, as in drawing along the body by the arms, the muscle serves as the chief bond of union between the shoulder and arm bones.

Triceps extensor cubiti.—Two heads of this muscle, outer and middle, are visible in the Plate.

The outer head, G, attached to the upper part of the back of the humerus, reaches nearly as high as the insertion of the teres minor, L, and is covered by the deltoid.

The middle or long head, F, is fixed to the inferior costa of the scapula close to the shoulder-joint. This part enters between the two teres muscles (over the major and under the minor), and divides into two the triangular space included by them. In front of the head, between it and the humerus, is a quadrangular interval, through which the posterior circumflex vessels and nerve turn from the axilla; and behind the head is an opening triangular in shape, which transmits the dorsal branch of the subscapular vessels.

A knowledge of the attachments of the muscles to the upper part of the humerus will be serviceable in counteracting in fracture of that bone the displacement of the fragments. In fracture of the neck of the bone near the teres minor insertion the upper end, into which the three dorsal scapular muscles are inserted, will be fixed in the glenoid hollow, and tilted rather outwards. Whilst the lower end will be inclined inwards towards the trunk by the latissimus dorsi, teres major, and pectoralis major, pulling in the direction of their fibres; and it will be finally carried upwards inside the upper fragment by the contraction of the muscles coming from the scapula to the humerus, viz., deltoid, coraco-brachialis, and triceps.

In an oblique fracture lower down (about opposite N on the deltoid) the relative position of the fragments to each other would be reversed. In that case the upper fragment will be drawn inwards towards the trunk by the latissimus, teres major, and pectoralis major; but though the lower end of the humerus will be elevated by the muscles descending from the scapula, as before said, it will be placed outside the upper end by the power of the deltoid muscle alone.

ARTERIES OF THE SHOULDER.

The shoulder possesses few vessels in comparison with some other parts. Two small arteries with their veins are met with in this region, and they are derived from the axillary trunk.

The *posterior circumflex artery*, a, one of the lowest branches of the axillary trunk (Plate i. *h*), appears between the humerus and the long head of the triceps; and winding forwards round the shaft of the humerus, it is distributed to the under surface of the deltoid muscle.

It supplies chiefly the deltoid, but offsets enter also the teres minor, and the long head of the triceps. Some branches are given to the head of the humerus, and anastomose in front with the anterior circumflex. A cutaneous offset descends to the integuments over the deltoid.

In the operation of amputation at the shoulder-joint the assistant follows the knife with his hands to seize the large axillary artery when it is divided, but he cannot compress at the same time the circumflex artery placed much farther back. This vessel pours out blood freely, and it may be secured first, provided the assistant controls the bleeding of the axillary trunk.

The *dorsal scapular artery, b,* is an offset of the subscapular branch of the axillary (Plate ii. *f*). Appearing through the triangular space behind the long head of the triceps, it bends round the edge of the scapula under the teres minor, and ramifies in the infra-spinal fossa.

As it is about to enter the fossa a branch is directed along the inferior border of the scapula, between the teres muscles, to which and the integuments it is distributed.

NERVE OF THE SHOULDER.

A large nerve from the brachial plexus ramifies under the deltoid muscle.

The *circumflex nerve*, 1, which is delineated in Plate i. 12, accompanies the posterior circumflex artery to the shoulder. Like the vessel it ends mostly in the deltoid muscle, supplying offsets to the fleshy fibres as it winds over the humerus.

Close to the border of the teres minor a considerable branch, 2, breaks up into offsets to the teres, the back of the deltoid, and the integuments covering the lower part of the deltoid muscle. In the natural position of the integuments the cutaneous branch would wind forwards over the muscle.

On the branch to the teres minor, 3, there is usually an enlargement of a reddish color and elongated form, which has been designated a "gangliform swelling." Before the nerve is disturbed that swelling lies close to the teres muscle.

In consequence of the loop made by the circumflex nerve under the head of the humerus, compression of it with impairment of function follows dislocation downwards of that bone. Paralysis of the deltoid muscle, with inability to raise the arm, will follow considerable disease or injury of the circumflex nerve.

DESCRIPTION OF PLATE VI.

The triceps muscle at the back of the arm and some of the shoulder muscles are here displayed. Whilst the Drawing was in progress the body was raised on blocks, and the arm was fastened over the side of the table.

To lay bare the triceps carry an incision along the back of the arm, and reflect the integuments and the deep fascia beyond the elbow. Generally the limb has been separated from the trunk before the student undertakes the dissection; in such case the triceps muscle may be made tense by a block beneath the elbow.

MUSCLES OF THE ARM AND SHOULDER.

On the back of the humerus lies the large triceps muscle, which extends the elbow-joint.

The shoulder muscles have been described with Plate v., and will require but little additional notice; the scapular muscles are more stretched in this than in the preceding Plate.

A. Rhomboideus major.
B. Latissimus dorsi.
C. Teres major.
D. Teres minor.
E. Infra-spinatus.
F. Deltoid muscle.
G. Inner head of the triceps.
H. Middle head of the triceps.
K. Outer head of the triceps.
L. Tendon of the triceps.
N. Fascia over the infra-spinatus.
O. Fascia of the arm.
P. Spine of the scapula.

The *triceps extensor cubiti* is undivided below, but is split into three processes of origin above, viz., the outer, inner, and middle heads.

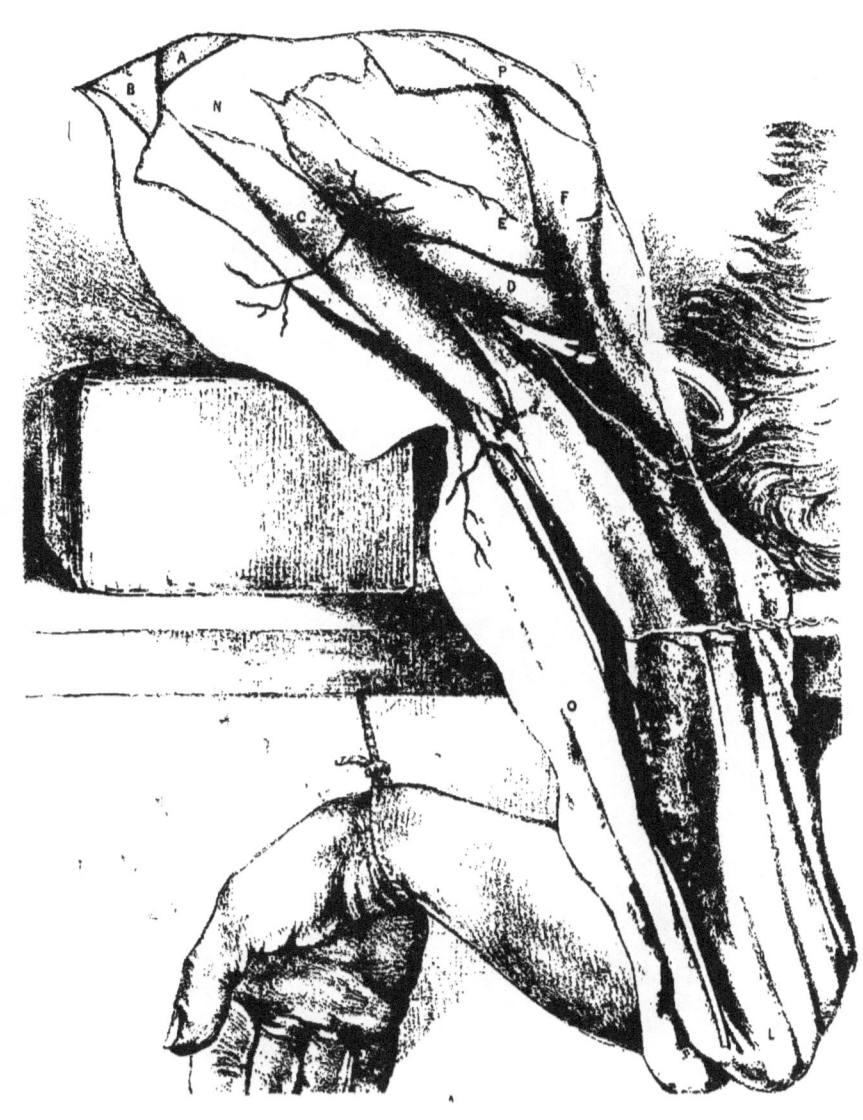

The outer head, K, is attached along the upper half of the posterior surface of the humerus, above the groove for the musculo-spiral nerve and its vessels, and reaches upwards nearly to the teres minor (Theile).* This attachment is represented in Plate vii.

The inner head, G, larger below than above, and concealed by the middle head, arises from the hinder surface of the humerus below the winding groove, extending laterally to the intermuscular septa, and upwards to the insertion of the teres major (Theile). See Plate vii. for its extensive origin.

The middle or long head, H, reaches the inferior or axillary border of the scapula, from which it takes origin for about an inch.

The outer and middle heads blend about the middle of the arm, but the inner one joins lower down. The muscle ends below in a wide, strong tendon, which receives deep fleshy fibres down to the elbow-joint, and is inserted into the end of the olecranon process of the ulna—a small bursa lying between the tendon and the tip of that piece of bone.

This muscle is represented in the thigh by the extensor muscle of the knee-joint. It is subcutaneous except above; and it is separated laterally from the muscles in front of the humerus by processes of fascia—the intermuscular septa. The long head lies between the teres muscles.

By the action of this muscle the elbow-joint will be extended; and supposing the limb removed from the body, it can be approximated to the trunk by the long head. But should the upper limb be fixed at a distance from the side, the muscle can assist in moving the trunk (through the scapula) towards the fixed arm, as in dragging the body forwards by a rope.

When the olecranon process of the ulna is detached by fracture, it is drawn upwards by the triceps, as far as the lower fleshy fibres of the muscle will allow, in the same manner as the upper fragment of the patella, in transverse fracture of that bone, is carried upwards by the extensor cruris. When replacing the displaced fragment force is not to be employed alone for the purpose of drawing it down towards the end of the ulna; but the interval is also to be diminished by moving backwards the shaft of the ulna by the extension of the elbow-joint.

In dislocation forwards of the humerus the olecranon becomes very prominent behind the elbow, and the tendon of the extensor muscle stands

---

* See a foot-note to the description of the triceps belonging to Plate VII.

out from that bone something like the tendo Achillis in the leg. Also some of the lower fleshy fibres will be broken through by the humerus being forced from the tendon.

By the action of the triceps, fracture of the lower end of the humerus near the elbow may be made to resemble the dislocation above noticed; for the lower end of the bone entering into the elbow-joint is forced upwards behind the rest of the shaft by the contracting muscle, and the olecranon is rendered more than usually prominent. But the nature of the injury may be made out by attention to the place of the olecranon:— in a dislocation this point of the bone projects much beyond, and is higher than the condyles of the humerus. but in fracture of the bone it is not more prominent with respect to the condyles than in the other limb, and it retains its usual position on a level with them.

Deltoid muscle, F.—At the origin of this muscle from the spine of the scapula it is tendinous behind, and blends with the fascia covering the infra-spinate muscle. The hinder part of the muscle has been turned forwards to allow a sight of the circumflex vessels and nerve beneath

Latissimus dorsi, B.—The muscle has been cut and thrown down as it crosses the angle of the scapula; the extent to which it covers that point of bone, and the rhomboideus major and teres major muscles, may be observed.

ARTERIES OF THE ARM AND SHOULDER.

The trunk of the brachial artery and some of its offsets are met with in the dissection of the back of the arm. Branches of the axillary artery are distributed to the shoulder.

a. Dorsal scapular artery.	*d.* Muscular branch of the brachial.
b. Circumflex artery.	*e.* Trunk of the brachial.
c. Muscular offset of the superior profunda artery.	*f.* Muscular branch of the brachial.
	g. Inferior profunda artery.

The *brachial artery*, *e*, is visible from behind where it lies inside the humerus, but it disappears below by passing in front of the arm bone. Contiguous to the upper part of the artery is the triceps muscle, viz., the middle head, H, and the inner head, G. A large companion vein (the continuation of the basilic) is placed on the inner side.

Close inside the artery is the ulnar nerve, 3; and intervening between it and the middle head of the triceps is the musculo-spiral nerve, 4.

Two *muscular offsets*, *d*, and *f*, enter the long head of the triceps.

The *upper profunda*, or the muscular artery of the back of the arm, is concealed by the middle head of the triceps; an offset, *c*, from it enters the outer head of that muscle. The distribution of this branch is represented in Plate vii.

Inferior profunda artery, g.—Winding backwards from the brachial (Plate iv.), it accompanies the ulnar nerve, 3, to the interval between the olecranon and the inner condyle, where it joins a branch of the ulnar artery.

The *dorsal scapular artery, a,* courses under the teres minor muscle, p. 53. Amongst the surrounding muscles supplied by it is the deltoid, to which it gives an offset: this was cut through in the dissection.

The position of the *posterior circumflex artery, b,* to the deltoid appears in this view of the parts. Some of its muscular offsets, and the branch to the integuments, are apparent.

NERVES OF THE SHOULDER AND ARM.

The nerves of the shoulder and back of the arm are branches of the brachial plexus, and have been partly represented in other Illustrations.

1. Circumflex nerve.
2. Offset of the musculo-spiral to the middle head of the triceps.
3. Ulnar nerve.
4. Musculo-spiral nerve.

Circumflex nerve, 1.—The anatomy of the trunk of the nerve can be studied in Plate v. Its cutaneous offset retains its natural place in this dissection.

The *musculo-spiral nerve*, 4, winds from the inner to the outer side of the limb between the humerus and the triceps muscle (Plate vii.). The figure, 2, marks an offset from it to the middle head of the triceps.

The *ulnar nerve*, 3, lies along the inner side of the arm as far as the elbow (Plate iv.). In the lower half of its course it is placed at the back of the limb, behind the inner intermuscular septum, and is partly concealed by fibres of the inner head of the triceps.

In excision of the articular ends of the bones of the elbow-joint through the triceps, the ulnar nerve is liable to be cut. To secure it from accident the nerve is dislodged from its hollow during the operation, and is moved to the front of the projecting inner condyle of the humerus. Temporary loss of the power of contraction in the muscles, and of feeling in the integuments of the inner part of the forearm and hand, follows division of the nerve; and this lost power would not be regained till the nerve structure has been repaired.

DESCRIPTION OF PLATE VII.

DISSECTION of the musculo-spiral nerve at the back of the arm, with its accompanying artery—the profunda.

Supposing the triceps denuded, as in Plate vi., the middle and outer heads are to be cut through after the manner shown in the Figure, to trace the nerve and its vessels. At the outer part of the muscle, a small branch of nerve and artery should be followed through the fleshy fibres to the anconeus muscle of the forearm.

MUSCLES OF THE ARM AND SHOULDER.

After the triceps has been divided in the way indicated, the attachment of the inner and outer heads to the humerus becomes evident.

The shoulder muscles have been displaced but little during the dissection, but they are shown on the stretch in consequence of the limb being placed in a hanging posture.

A. Lower end of the long head of the triceps cut through.
B. Upper end of the long head of the triceps.
C. Outer head of the triceps.
D. F. Inner head of the triceps.
E. The nerve and vessels to the anconeus.
G. Anconeus muscle.
H. Supinator longus muscle.
K. Teres minor muscle.
L. Infra-spinatus muscle.
M. Latissimus dorsi muscle.
N. Teres major muscle.
P. Deltoid muscle.
Q. Outer condyle of the humerus.
R. Olecranon process.
S. Fascia of the forearm reflected.

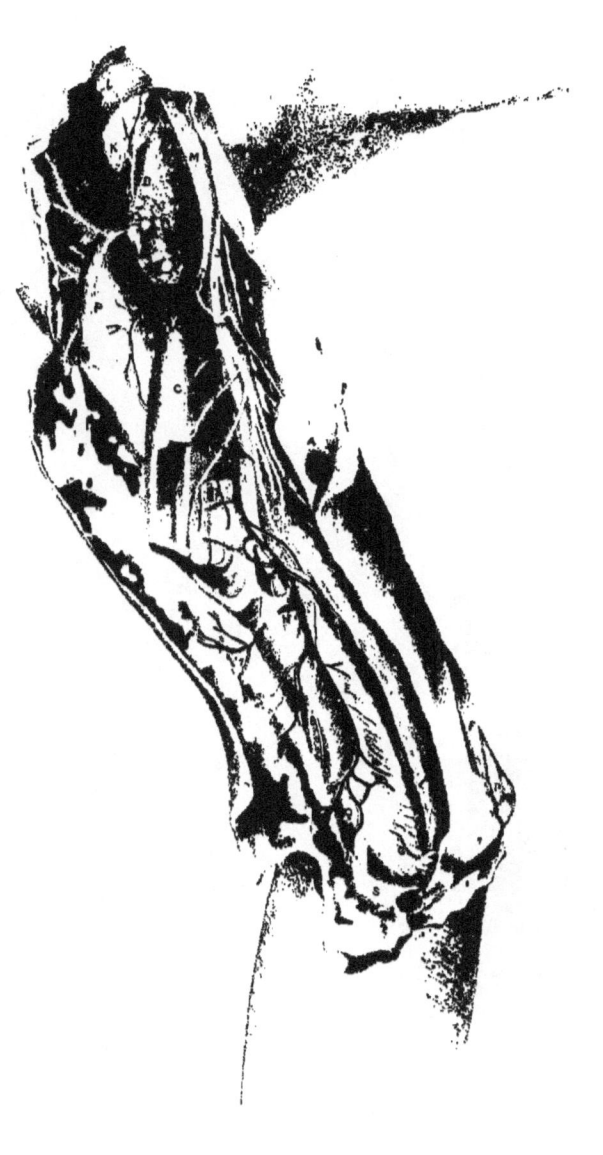

Triceps extensor brachii.—The superficial view of this muscle, and the attachment of the middle head can be seen in Plate vi. Only the origin of the outer and inner heads will be noticed below.

The outer head, C, arises at the back of the humerus above the groove in the bone, which lodges the musculo-spiral nerve and the vessels; it narrows above as it ends near the insertion of the teres minor.

The inner head, D, and F, arises from the back of the humerus below the winding groove, reaching upwards by a pointed part as high as the teres major muscle—sometimes to the upper, and sometimes the lower border. This head is wide below, and reaches laterally to the intermuscular septa, from which fibres take origin.*

Subanconeus.—Some of the deepest fibres of the triceps near the elbow terminate in the capsule of the joint, like fibres of the extensor of the knee, and are said to constitute a separate muscle, to which the name subanconeus has been applied; but I have not observed such isolated and distinct muscular bands as Anatomists describe.

Supinator longus muscle, H.—Covered by the fascia of the limb the muscle is fixed to the outer condyloid ridge of the humerus, as high as the groove before referred to. This muscle and the extensor carpi radialis longus occupy the ridge—the former reaching the upper two thirds, and the latter, the lower third. Above the upper border of the supinator the musculo-spiral nerve and vessels are directed forwards.

The group of shoulder muscles is strained by the weight of the arm, as in Plate vi.

Naturally the *teres minor muscle,* K, is not so covered by the long head of the triceps; but as this head was cut, and the limb hanging, its upper end, B, was pushed back by the latissimus dorsi arching in front.

The *latissimus dorsi,* M, and *teres major,* N, are stretched as they descend to the humerus. Only the upper edge of the teres appears; below and in front they are partly blended by tendinous fibres.

By the weight of the limb the *deltoid muscle,* P, is made to look flatter than it is usually.

---

* This statement of the origin of the inner and outer heads of the triceps differs much from the common Anatomical description. It contains the view of Theile, and has the merit of being more accurate. The original account is given in Müller's Archiv für Anatomie, etc., for 1839, p. 420—"Ueber den Triceps brachii und den flexor digitorum sublimis des Menschen."

VESSELS OF THE BACK OF THE ARM.

The ramifications of the superior profunda artery through the triceps, and its origin from the brachial trunk, are contained in this region. Some small branches of the circumflex artery appear behind the border of the deltoid.

- a. Brachial artery.
- b. Basilic vein becoming axillary.
- c. Superior profunda artery.
- d. Offset of the profunda to the front of the arm, with the musculo-spiral nerve.
- f. Branch of artery along the outer intermuscular septum.
- g. Inosculating artery from the recurrent radial.
- h. Anastomotic branch from the interosseous recurrent artery.
- k. Muscular branch of the artery to the triceps and anconeus.
- l. Branch of artery to the teres minor from the posterior circumflex.
- m. and n. Cutaneous and muscular offsets of the posterior circumflex artery.

Brachial artery, a.—The anatomy of the brachial trunk issuing from the armpit has been described with Plate vi., p. 56. The following large muscular offset springs from this part of the artery.

The *superior profunda* branch, c, is the nutritive and anastomotic vessel of the back of the arm, and corresponds with the profunda artery of the femoral trunk in the thigh. Springing from the brachial, near the axilla, it is the largest offset of that vessel, and winds behind the humerus in the hollow separating the inner and outer heads of the triceps, as far as the outer side of the limb, where it ends in muscular, anastomotic, and cutaneous offsets.

The *muscular branches* supply the three heads of the triceps, viz., the long head, A, the external, C, and the internal, D. A second artery enters the long head from the brachial trunk.

The *anastomotic offsets*, three in number, spring from the end of the profunda. One, b, variable in size, accompanies the musculo-spiral nerve to the front of the arm, and communicates with the radial recurrent. A second, f, runs on the intermuscular septum to the outer condyle, and anastomoses with a branch, g, of the radial recurrent, and with a branch, h, of the recurrent interosseous; and the third artery, k, descends in the triceps to the hollow between the outer condyle and the olecranon, and entering the anconeus, G, joins in the last muscle with the recurrent of the interosseous.

The *cutaneous offsets*, two or three in number, pass out with nerves to the integuments, and are derived, for the most part, from the branch, *f*.

The *posterior circumflex artery* enters under the deltoid muscle (Plate v.). From the part of the artery now visible spring the branch to the teres minor, *l*, and the offsets to the integuments and the deltoid, *m*, and *n*.

The usual *companion veins* run with the arteries, though they are not included in the Plate; those with the profunda artery join a brachial vein; and the circumflex veins open into the axillary trunk.

NERVES OF THE BACK OF THE ARM.

The nerves correspond in the main with the vessels. With the profunda is situate the large musculo-spiral nerve, distributing branches to the triceps and the integuments; and by the side of the circumflex artery lies the muscular nerve of the same name, which ends in the deltoid.

1. Musculo-spiral nerve.
2, 2. Ulnar nerve.
3. Offset to the long head of the triceps.
4. Offset to the inner head of the triceps.
† † † Three branches to the outer head of the triceps.
6. Branch to the anconeus.
7. Upper external cutaneous of the musculo-spiral.
8. Lower external cutaneous of the musculo-spiral.
9. Nerve to the teres minor.
10. Cutaneous branch of the circumflex nerve.

The *musculo-spiral nerve*, 1, begins in the brachial plexus (Plate i. 13); and, reaching the digits, supplies the extensor and supinator muscles on the back of the arm and forearm, together with some of the integuments.

In the arm the trunk winds behind the humerus from the inner to the outer side, and divides at the outer condyle into two—radial and posterior interosseous nerves (Plate xii.). The nerve lies in the groove of the humerus between the inner and outer heads of the triceps, and turns to the front of the arm above the supinator longus muscle, H.

Offsets of the part of the nerve now dissected supply the extensors of the elbow-joint and the teguments.

Muscular branches enter the heads of the triceps. One, 3, belongs to the parts, A, and B, of the long head; others, 4, and 6, supply the inner head, D; and three † † † enter the outer head, C. To the inner and long heads some branches are furnished by the trunk of the nerve in the axilla (Plate iv. 2).

The branch, 6, of the anconeus is very slender, and is contained in the triceps.

Cutaneous nerves.—Two external cutaneous appear with superficial arteries on the outside of the limb; the upper one, 7, smaller than the other, reaches in the integuments of the arm as far as the elbow; and the lower nerve, 8, is continued beyond the elbow, on the back of the forearm, nearly to the wrist.

Whilst the musculo-spiral is contained in the axilla it furnishes an internal cutaneous nerve to the inner and hinder parts of the arm (Plate iv. 1).

If the musculo-spiral nerve is cut across, or its action much impaired by disease, the extensor muscles of the elbow-joint, amongst others, would be incapable of contracting; and the elbow would therefore be bent by the flexors which, being uncontrolled by their antagonists, would carry forwards the forearm bones.

Ulnar nerve, 2, 2.—The upper and lower parts of this nerve come into view in the Figure. The whole course of the nerve appears in Plate iv.

Circumflex nerve of the shoulder.—The trunk of the nerve is noticed in the description of Plate v. Two offsets, viz., one marked, 9, for the teres muscle, and another, 10, for the integuments, appear behind the deltoid muscle.

DESCRIPTION OF PLATE VIII.

The dissection of the muscles, vessels, and nerves of the front of the forearm, with their connections undisturbed by the reflection of the deep fascia, is here displayed.

All the superficial coverings of the limb may be removed at once by an incision along the front of the forearm, met by a cross-cut a little above the elbow, and by another rather below the wrist. But a more profitable dissection may be made by examining, and afterwards removing in successive layers, the skin, the subcutaneous fat with its vessels and nerves, and the deep fascia.

SUPERFICIAL MUSCLES OF THE FOREARM.

Inside the line of the brachial and radial arteries, b and f, lies a group of muscles which act as flexors and pronators; and outside the vessels is a mass of muscles consisting of extensors and supinators, antagonists of the former set.

The inner group is divided into two strata, superficial and deep. Five muscles belong to the superficial layer: of these one is a pronator of the hand, and the others are flexors of the wrist and fingers.

A. Biceps flexor brachii.
B. Brachialis anticus.
C. Pronator teres.
D. Palmaris longus.
E. Flexor carpi radialis.
F. Flexor digitorum sublimis.
G. Flexor carpi ulnaris.
H. Flexor longus pollicis.
L. Supinator longus.
N. Palmaris brevis.

P. Extensor carpi radialis longior.
S. Extensor ossis metacarpi pollicis; close alongside is the tendon of the extensor primi internodii pollicis.
† Inner intermuscular septum of the arm.
* Slip of fascia connecting the tendon of the flexor carpi ulnaris with the annular ligament.

Pronator radii teres, C, the first muscle of the inner group, arises in part from the common origin;* from the condyloid ridge of the humerus; and from the coronoid process of the ulna by a separate slip (Plate ix.). Below it is inserted into the middle of the radius beneath the supinator longus, L.

By its outer edge the muscle bounds the hollow in front of the elbow, and by the other it touches the flexor carpi radialis. Near the insertion the radial vessels rest on it.

When the pronator first contracts it will roll the radius over the ulna, pronating the hand; and acting still more, it will bend the elbow-joint over which it passes.

The *flexor carpi radialis*, E, having the common origin, is continued through a groove in the os trapezium to be inserted chiefly into the base of the metacarpal bone of the index finger, but also by a slip into the metacarpal bone of the middle finger.

The tendon of the muscle is prominent below outside the middle line of the forearm, and bounds internally a surface-depression over the radius which contains the radial artery; it may be taken as the guide to that vessel.

After bending the wrist, the muscle will approximate the forearm to the arm.

The *palmaris longus*, D, has the common origin between the preceding and the flexor carpi ulnaris, G; and its tendon piercing the aponeurosis of the limb near the wrist, ends in the fascia of the palm of the hand, after sending a slip to join the short muscles of the thumb. This muscle may be absent.

It renders tense the palmar fascia, and assists in bending the elbow and wrist.

The *flexor carpi ulnaris*, G, is the most internal muscle of the set. Attached to the inner condyle of the humerus, where it blends with the other muscles, and to the posterior ridge of the ulna by an aponeurosis, it is inserted into the pisiform bone, and joins by offsets the annular ligament of the wrist and the muscles of the little finger.

The outer edge of the muscle corresponds with a line from the pisiform

---

* Most of the superficial muscles of the forearm, on both the front and back, have a common origin from the fascia of the limb, and from a strong fibrous process (tendon of origin) which is attached to the condyle of the humerus in each case, and sends pieces between the muscles.

bone to the inner condyle of the humerus, and there is a surface-groove in the lower third of the forearm over that edge. The muscle conceals below the ulnar vessels and nerve.

Its main action is expressed by its name, but it serves also as a flexor of the elbow-joint.

The *flexor digitorum sublimis*, F, is the deepest of the muscles of the superficial layer. It is attached by its thin outer edge to the upper three fourths of the shaft of the radius; higher still, to the inner side of the coronoid process of the ulna; and finally to the lateral ligament of the elbow-joint, and the common tendon of origin of the other muscles. It ends below in four tendons for the fingers, which cross the hand, and are inserted into the middle phalanges (Plate x.).

The extent of attachment to the radius, and the position of the radial vessels to it may be noticed in the Drawing. Only two tendons appear on the surface, viz., those of the middle and ring fingers. Issuing beneath the lower border is the median nerve, 2.

Besides bending the phalanges, the muscle will contribute to flex the wrist and elbow.

Above the elbow are the flexors of that joint, viz., *biceps* and *brachialis anticus*. The first is inserted into the radius and the other into the ulna; and when they contract they carry forwards those bones over the end of the humerus.

After fracture of the olecranon process of the ulna—the part limiting the movements and giving security to the joint—the elbow is bent because these two muscles are stronger than the extensor muscles behind (the triceps being useless).

In dislocation of the humerus on the front of the ulna and radius, the flexor muscles give the bent state to the limb. Being greatly stretched, especially the brachialis, by the large projecting end of the humerus, they contract powerfully; and the forearm is carried forwards as much as it can be to relax the tense state of the muscular fibres.

The extensors and supinators on the outer side of the limb - sected only in part: they will be described --
They are divisible, like the muscle
layer. Only one of them will be 1

The *supinator longus*, L (brach   most anterior
and the longest of the external group   arises from the upper two thirds
of the condyloid ridge of the humerus in front of the outer intermuscular

5

septum (Plate xii.); and it is inserted into the lower end of the radius, close to the styloid process.

Covered at its origin and insertion by other muscles, it forms part of the outer swell of the forearm, and limits externally the hollow in front of the elbow-joint. It rests upon the long radial extensor of the wrist, P, and covers the radial artery in the upper half of the forearm. At its insertion it is crossed by the extensor muscles of the thumb, S.

Its chief office is to bend the elbow-joint. But it will become a supinator when the hand is quite prone; and, when the hand is strongly supinated, it is said to bring the same into the prone position.

HOLLOW IN FRONT OF THE ELBOW.

This intermuscular space between the inner and outer groups of muscles is represented in the lower limb by the ham. It contains the chief vessel of the arm and the companion nerve; and by its position on the aspect of the limb to which the joint is bent, greater freedom of movement forwards is permitted.

The interval is somewhat triangular in form, as seen on the surface, and has the following boundaries:—stretching over it is the aponeurosis of the limb joined by an offset from the biceps tendon, with the integuments and the superficial veins and nerves (Plate iii.); and covering the underlying bones are the brachialis anticus, B, and supinator brevis. Externally is placed the supinator longus, L, and internally the pronator teres, C; the fibres of the former being nearly straight in the forearm, and those of the latter slanting downwards and outwards. The base is turned towards the arm; and the apex points forwards in the forearm.

Contained in the hollow is the tendon of the biceps, with vessels, nerves, fat, and lymphatics; and their position in it is as follows:—

On the outer side is the biceps muscle, A, whose tendon dips into the space to reach its insertion into the radius.

The brachial artery, b, lies close inside the biceps, and divides, opposite the "neck of the radius" (Quain), into the two arteries of the forearm, which are directed forwards through the space, the radial being superficial and the ulnar deep in its position. Venæ comites entwine around the arterial trunks. Small arteries are found in the space. Thus in the outer part the recurrent of the radial artery (Plate xii. 3) is directed transversely to the supinator longus; and in the inner part of the hollow, offsets of the anastomotic artery, *a*, descend beneath the pronator teres to join the anterior recurrent of the ulnar artery. Other cutaneous offsets, *c* and *d*, come forwards to the integuments from the brachial and the radial artery.

Inside the artery, and separated from it by a slight interval, which increases below to a quarter or half an inch, comes the median nerve, 2. At this spot the nerve supplies small offsets to the inner group of muscles of the forearm. Underneath the supinator longus, and therefore outside the superficial limits of the space, the musculo-spiral nerve may be found dividing in front of the condyle of the humerus into radial and posterior interosseous branches.

Loose fat fills the hollow, supporting the vessels and nerves, and extends into the forearm along the bloodvessels; and blood effused beneath the fascia finds its way along the same channels.

A few lympathic glands with their connecting vessels accompany the arteries—two or three lying on the sides of the brachial, and one below its point of splitting.

From the lax condition of the parts surrounding the brachial artery pressure applied to the vessel, when wounded, should be firm and graduated. The limb too should be kept still; for when the elbow is moved much the vessel may slip away from the compressing pad, and blood may be effused beneath the fascia.

ARTERIES OF THE FOREARM.

Two chief vessels occupy the front of the forearm, and these spring from the division of the brachial trunk. They are named radial and ulnar from their position in the limb; and both reach the palm of the hand, where they form arches and supply branches to the fingers. Both are placed deeper near the elbow than at the wrist.

a. Anastomotic branch of the brachial trunk.
b. End of the brachial artery.
c, *d*. Unnamed cutaneous offsets: the former from the brachial, and the latter from the radial artery.

f. Radial artery.
g. Superficial volar branch.
h. Ulnar artery.
n. Cutaneous median vein, joining a deep companion vein.

The *radial artery, f,* is the more external of the two bloodvessels in the forearm, and inclines from the bifurcation of the brachial trunk to the lower end of the radius; it then winds to the back of the wrist below the radius, and enters the hand. The part from the wrist onwards will be included in other dissections (Plates x. and xi.). A line from the centre of the elbow-joint to the styloid process of the radius will mark the course of the vessel on the front of the forearm.

In the *upper half* of the forearm the artery is concealed by the supinator longus;* and it rests in succession on the supinator brevis, pronator teres, C, and flexor sublimis digitorum, F.

Venæ comites lie on the sides of the artery. But no nerve is in contact with it—the radial being placed too far out.

This part of the artery may be superficial to the long supinator, lying even in the integuments, when there is an unusual origin from the brachial.

In the *lower half* of the forearm the vessel is not covered by muscle, but is contained in a hollow between the tendons of the flexor carpi radialis, E, and supinator longus, L. Only the common teguments cover the vessel here. It is supported by part of the flexor sublimis, F, flexor longus pollicis, II, and lower down by the pronator quadratus and the end of the radius.

The usual veins surround the artery. The radial nerve, 3, is at some little distance outside the vessel, and becomes cutaneous behind the tendon of the supinator longus.

The *offsets* of the radial artery are for the most part small, but near the elbow and wrist they acquire greater size. No one is large enough usually to interfere with the placing a ligature on the trunk.

Ligature of the radial artery.—In the upper half the vessel would not require to be tied in the living body unless it was wounded. In seeking it amongst the tissues infiltrated with blood the supinator longus, and the line of the vessel, will serve as material aids to the surgeon.

In the lower third of the forearm, the radial may be secured for a

---

* In Anatomical Plates the radial artery is usually delineated with the supinator longus removed from it, as if the vessel was uncovered by muscle in the upper half of the forearm. In this Plate the muscle is shown covering the artery, as it exists before it is displaced, to impress upon the memory the fact that where the radial is so protected it cannot be easily injured.

wound in the palm of the hand. With a cut about one inch and a half long the integuments and superficial veins and nerves are to be divided in the line of the vessel. The fascia may be carefully cut for the same extent. After the sheath has been opened and separated from its contents in the usual way, the aneurism needle may be carried round the artery.

As this part of the radial is so superficial the student, when first practising the operation, cuts oftentimes, not only the coverings of the limb, but also the artery.

If the vessel is tied for a wound near the wrist two ligatures should be applied, although the size is so small, on account of the free communication of the radial with the ulnar artery in the palm of the hand.

Branches of the artery.—Small unnamed muscular and cutaneous offsets leave the trunk of the artery at intervals; and larger named branches arise near the beginning and ending.

The *recurrent radial* ascends under cover of the supinator longus, and anastomoses on the outer part of the elbow with the superior profunda (Plate xii.): it supplies some of the outer group of muscles.

The *superficial volar branch*, *g*, descends to the hand across or through the short muscles of the thumb. When small, it ends in those muscles (Quain); and when larger, it joins the superficial palmar arch (Plate x.). With this vessel of very unusual size a wound of it might require it to be tied.

The *anterior carpal branch* (Plate ix. *d*), which is generally so small as not to deserve notice, arises near the wrist, and is lost on the carpus.

Muscular and cutaneous branches arise at tolerably regular intervals. One to the integuments is marked by *c*. From a muscular branch near the wrist a twig entered the median nerve.

The *ulnar artery*, *h*, is concealed almost entirely by muscles whilst it is in the forearm, only a small part near the wrist being visible before the natural position of the flexor carpi ulnaris has been disturbed.* And the part of the artery, which is represented, appears smaller than it is com-

---

* In Plates of the vessels of the forearm, where the ulnar artery is laid bare to view in the lower third or more, the flexor carpi ulnaris has been drawn aside in the dissection. This rather deep condition of the artery should be kept in mind in any attempt to put a ligature on it.

monly, in consequence of being partly covered by the venæ comites. The course and the branches of the artery were shown in Plate ix.

NERVES OF THE FOREARM.

Three nerves, viz., median, ulnar, and radial, are visible each for a short distance in this dissection of the superficial muscles of the forearm.

1. Cutaneous part of the musculo-cutaneous, named external cutaneous of the forearm.
2, 2. Median nerve.
3. Radial nerve.
4. Cutaneous palmar branch of the median nerve.
5. Palmar part of the ulnar nerve.

The *median nerve*, 2, is superficial for two inches above the wrist, and is placed on the outer side of the tendons of the flexor sublimis. As it passes through the forearm it lies beneath the superficial flexors. From the forearm it is continued to the hand beneath the annular ligament. The following offset arises from this part of the nerve.

The *cutaneous palmar branch*, 4, pierces the deep fascia near the wrist, and crosses over the annular ligament to end in the integuments of the ball of the thumb and palm of the hand; at its ending it communicates with the ulnar nerve.

The *radial nerve*, 3, is a tegumentary branch of the musculo-spiral (Plate xii. 2), and ends on the back of the hand. Becoming superficial behind the tendon of the supinator longus, it terminates in the teguments of the back of the thumb, of the next two digits, and sometimes of half the ring finger.

The *ulnar nerve*, 5, enters the palm of the hand over the annular ligament; its termination is given in Plate x. This is the only part of the nerve which comes into sight in the forearm before the flexor carpi ulnaris has been turned aside; and it is partly concealed by the ulnar vessels.

DESCRIPTION OF PLATE IX.

This Plate represents the dissection of the deep muscles on the front of the forearm, with the vessels and nerves in contact with them.

To make ready the dissection cut through near the humerus and remove the inner group of the superficial muscles, seen in Plate viii., except the pronator teres on the outside, and the flexor carpi ulnaris on the inside; then draw upwards the pronator, and inwards slightly the flexor ulnaris from the ulnar vessels. The small veins with the branches of the arteries have been taken away.

DEEP MUSCLES OF THE FOREARM.

The deep muscles are three in number: two flex the digits, and one pronates the radius. One, a flexor of the thumb, lies on the radius; and the other large muscle, covering the ulnar, is the common flexor of the fingers. The pronator is placed beneath the other two near the wrist.

A. Lower end of the biceps.
B. Brachialis anticus.
C. Supinator longus.
D. Pronator teres.
F. Conjoined palmaris longus and flexor carpi radialis, cut, and, turned aside.
G. Flexor carpi ulnaris.
H. Supinator brevis.
J. Cut end of the flexor sublimis.
K. Flexor longus pollicis.
L. Flexor profundus digitorum.
N. Slip of flexor longus pollicis.
O. Extensor ossis metacarpi pollicis.
P. Pronator quadratus muscle.
Q. Tendons of flexor sublimis, cut.
R. Tendon of flexor carpi radialis.
X. Anterior annular ligament.
† Internal intermuscular septum of the arm.

The *flexor longus pollicis*, K, arises from the upper three fourths of the anterior surface of the shaft of the radius; from the contiguous interosseous membrane; and sometimes by a round slip, N, from the inner part of the coronoid process of the ulna. Its tendon passes beneath the annular ligament, X, and is conveyed along the thumb by a fibrous sheath to be inserted into the last phalanx.

Most of the muscle is covered by the flexor sublimis, but part of it below is in contact with the radial artery where the pulse is felt. Between the upper attachments of this muscle and the supinator brevis, II, to the radius, is a narrow slip of the bone from which the flexor sublimis digitorum arises.

The muscle bends the phalanges of the thumb, and brings the metacarpal bone towards the palm of the hand. It will flex the wrist after the digit.

Flexor profundus digitorum, L (perforans). It arises from the anterior and inner surfaces of the shaft of the ulna as low as the pronator quadratus; and other fibres spring from the membranes outside and inside the bony attachment, viz., from the interosseous membrane externally, and from an aponeurosis common to this muscle and the flexor carpi ulnaris internally. The fleshy fibres end in tendons which are united together above the wrist, only the most external being separate; and these, passing beneath the annular ligament, X, and across the hand, are inserted into the last phalanges of the fingers. See Plate x.

On the sides of the muscle are the flexor longus pollicis, K, and flexor carpi ulnaris, G. On it rest the ulnar vessels, and the ulnar and median nerves.

This muscle bends the last phalanx of each finger; and continuing its action it will aid in flexing the other phalanges and the wrist.

The *pronator quadratus*, P, lies beneath the preceding, and covers the lower ends of the bones of the forearm for about two inches, though more of the ulna than of the radius. Scarcely any part of the muscle is seen, but the interosseous nerve and artery pass beneath its upper edge, marking its extent upwards.

It is covered by the other two muscles of the deep layer, and the radial vessels touch the outer edge, near the wrist.

It acts on the radius, moving the lower end round the ulna so as to put down the palm of the hand.

Movement of the radius.—The rotatory motion of the hand is due to the movement forwards and backwards of the lower end of the radius over the ulna. When that bone is brought forwards the palm of the hand is placed down, or the limb is pronated; and when the bone is moved back the dorsum of the hand is turned towards the ground, and the member is supinated. The pronator muscles are in front, and passing from the inner side of the limb, draw forwards the radius; while the supinators, which turn back the bone, are placed on both the front and hinder part of the limb. The action of the supinators will be given with the description of Plate xii.

Two pronators are connected with the radius;—one, pronator teres of the superficial layer, being attached about midway between the ends; and the other, pronator quadratus, of the deep layer, is fixed into the lower part. Both are therefore inserted below the upper half of the bone; and during their contraction the lower end of the radius is moved over the ulna—the upper end not changing its position to that bone, but rotating in its band like a wheel. And as the active supinators (supinator brevis and biceps) are fixed to the upper part of the radius, their influence on the lower end is neutralized as soon as the bone is broken through at or near the middle; so that the lower fragment can be then moved forwards without obstacle by the action of the pronators.

Fracture of the radius near or below the middle is attended by pronation of the hand, and by displacement of the lower fragment, in consequence of the action of one or both of the pronators, and of the weight of the hand articulated to the radius. But the upper fragment of the broken bone does not change its place; it remains on the outer side of the ulna, though tilted away from that bone by the action of the supinators. Readjustment of the displaced lower fragment will be made by supinating the hand, for this movement carries back at the same time the lower end of the broken radius into contact with the upper. Future displacement of the lower fragment will be prevented if the weight of the hand is taken off by fixing the forearm and hand with splints in a position midway between pronation and supination, so that the thumb shall be in a line with the upper part of the radius, and the palm of the hand shall be turned to the chest.

Should the lower fragment not be brought well into line with the upper by the position of the forearm above-said, it will be necessary to place the hand quite supine (the palm of the hand looking directly upwards), and to fix it with splints in that posture, as was recommended by Mr. Lonsdale.*

In *fracture of the shafts of both bones* of the forearm, the lower ends, as in fracture of the radius, depart from the line of the upper ends, being

---

* "Fracture of the Forearm." By Edward Lonsdale. Medical Gazette, 1832, p. 910.

dragged away by the weight of the hand. They have further a tendency to approximate across the interosseous space, and will therefore be easily made to touch by any constriction, such as a bandage round the limb.

By supinating the hand in the manner described for fracture of the radius, the lower displaced ends will be brought to the upper fixed parts of the bones. And with the view of keeping apart the bones, gentle pressure with a narrow graduated pad is sometimes employed along the front and back of the forearm in a line with the interval between them. Pressure by means of a bandage is not to be made on the member, lest the broken ends be brought together, and the movements of the radius be lost by this bone blending with the ulna in the process of union. Re-displacement of the apposited ends may be prevented by splints reaching from the elbow to the fingers.

ARTERIES OF THE FOREARM.

Both radial and ulnar arteries are laid bare in the dissection, but the anatomy only of the ulnar and its branches will be now given. For a short distance above the elbow-joint the brachial trunk is shown.

- *a*. Brachial artery.
- *b*. Radial artery.
- *c*. Ulnar artery.
- *d*. Anterior carpal branch of the radial trunk.
- *e*. Superficial volar branch.
- *g*. Posterior ulnar recurrent branch.
- *k*. Anterior interosseous.
- *n*. Median artery.

The *ulnar artery*, *c*, tends to the inner side of the limb, and enters the palm of the hand in front (Plate x.). It keeps the name "ulnar" from the bifurcation of the brachial trunk to the lower border of the annular ligament, X.

The artery has a curved course in the forearm, being directed inwards in the upper part, but taking a straight direction at the lower part. A line on the surface, to mark the straight part of the artery, should be drawn from the inner condyle of the humerus to the inner side of the pisiform bone. The vessel is covered by muscles in the upper half of the forearm, but becomes more superficial below.

In the *deep part* of its course, viz., between the origin and the meeting with the flexor carpi ulnaris, G, the artery is curved with the convexity upwards. It is covered by the superficial layer of muscles except the flexor carpi ulnaris; and it rests firstly on the lower part of the brachialis anticus, B, and afterwards on the flexor profundus digitorum, L.

Companion veins are ranged on its sides, with communicating branches over it.

The median nerve, 1, is placed inside the ulnar artery for about an inch; it then crosses over, and leaves that vessel in the forearm. The ulnar nerve, 3, approaches the artery about half way between the wrist and elbow-joints, from which point it is situate inside, and close to the vessel.

The *lower half* of the artery lies along the flexor carpi ulnaris, G, by which it is overlapped (Plate viii.); and it is therefore more deeply placed than the corresponding part of the radial bloodvessel. On its outer side are the tendons of the flexor sublimis digitorum, F (Plate viii.), and it lies on the flexor digitorum profundus, L.

The companion veins join together freely over the artery, and the ulnar nerve, 3, is in contact with it on the inner side. Filaments of the palmar cutaneous branch, 6, of the ulnar nerve entwine around the vessel.

As the artery rests on the annular ligament of the wrist, it is very near the pisiform bone; it is crossed by a slip from the flexor carpi ulnaris to the annular ligament, and is concealed by some fleshy bundles of the palmaris brevis muscle (Plate viii.). The nerve, still internal, intervenes between the bone and the bloodvessel.

All the offsets of the lower part are too small to be considered of moment in ligature of the artery.

Ligature of the artery at its lower fourth, which is sometimes practised for a wound of the trunk in the palm of the hand, is a simple operation; and an inspection of Plate viii. will render more intelligible the following remarks.

Drawing back the inner part of the hand so as to stretch and depress the tendon of the flexor carpi ulnaris, make a cut about two inches long in the hollow observable on the surface, and carry it through the integuments and the deep fascia down to the flexor tendon. By bending now the wrist, the tendon will be relaxed, and can be moved aside. Under the muscle, but covered by a deeper layer of fascia, which is to be divided, the vessels and nerve will appear—the nerve being internal and serving as the deep guide to the artery.

When the sheath has been opened, and the artery detached from it and the surrounding veins, the needle carrying the ligature can be passed easily under the vessel.

In tying the vessel for a wound near the wrist two ligatures are to be used, as in the radial artery, because blood may be poured out above and below.

Branches. Named offsets arise near the large joints of the wrist and elbow, and smaller muscular branches leave the trunk at short intervals.

The *posterior recurrent branch, g,* is continued beneath the superficial layer of muscles to the space between the inner condyle of the humerus and the olecranon process, where it supplies the joint, and communicates with the inferior profunda and anastomotic branches (Plate iv.).

Near the beginning, a small branch, *anterior ulnar recurrent,* ascends under the pronator teres to join the anastomotic branch.

The *interosseous artery* arises near the preceding, and divides into two, anterior and posterior, for the front and back of the limb. The posterior is seen in Plate xii.

The *anterior interosseous, k,* runs on the interosseous membrane between the two deep flexors as far as the pronator quadratus, P, where it passes through the membrane to end on the back of the wrist (Plate xii.): as the artery leaves the front of the limb it sends a branch on the interosseous membrane to the fore part of the wrist.

It supplies branches to the deep muscles. Another offset *median, n,* ends in the median nerve and the flexor sublimis muscle: sometimes this last branch is large, and is continued with the nerve to join the palmar arch in the hand.

A *metacarpal branch* proceeds along the inner edge of the metacarpal bone of the little finger, on which it ends.

A small *anterior carpal branch* takes origin opposite the lower edge of the pronator quadratus: it joins the corresponding branch of the radial artery.

Some *cutaneous offsets* pass forwards to the integuments at the outer edge of the flexor carpi ulnaris: three of these may be observed in Plate viii.

NERVES OF THE FOREARM.

The median and ulnar nerves supply the muscles on the front of the forearm, whilst the integuments receive nerves mostly from other trunks. The two have a marked difference in position when entering and leaving the forearm: thus above, the median is superficial in front of, and the ulnar is behind the elbow; but below, the median is deeply placed beneath the annular ligament, whilst the ulnar passes over the ligament.

1. Trunk of the median nerve.
2. Anterior interosseous branch.
3. Ulnar nerve.
4. Branches of ulnar nerve to flexor carpi ulnaris muscle.
5. Branch of ulnar nerve to flexor digitorum profundus.
6. Cutaneous palmar branch of the ulnar.
7. Palmar cutaneous nerve of the median.

The *median nerve*, 1, courses between the superficial and deep layers of muscles, till about two inches above the wrist where it approaches the surface (Plate viii.). It distributes nerves to all the superficial muscles except the flexor carpi ulnaris, and offsets of its interosseous branch supply the deep muscles.

Muscular offsets may be seen entering the pronator teres, D, the palmaris longus, and flexor carpi radialis, F, and the flexor sublimis, J.

The *anterior interosseous branch*, 2, runs on the front of the interosseous membrane, with the artery of the same name, between or in the fibres of the flexors of the digits, and ends below in the pronator quadratus, P. It supplies the outer half of the flexor digitorum profundus, and the whole of each of the other two deep muscles, viz., flexor pollicis, and pronator quadratus.

The *cutaneous palmar branch*, 7, arises near the wrist: it is described with Plate viii.

The *ulnar nerve*, 3, is directed through the forearm along the flexor carpi ulnaris muscle, in the position of a line from the inner condyle of the humerus to the pisiform bone. Branches are supplied to one muscle and a half.

Articular filaments. Behind the elbow one or two slender twigs are furnished to the joint.

Muscular offsets. One or two nerves, 4, enter the flexor carpi ulnaris; and one, 5, belongs to the inner half of the flexor digitorum profundus.

The *palmar cutaneous branch*, 6, is conveyed along the lower half or third of the ulnar artery to the integuments of the palm of the hand: it sends offsets around the artery, and communicates with the palmar branch of the median nerve at its ending.

DESCRIPTION OF PLATE X.

Views of the two dissections of the palm of the hand, which are needed to lay bare the superficial and deep muscles, vessels, and nerves.

Figure 1.

In the left-hand Figure the superficial palmar arch of the ulnar artery, with its offsets, also the nerves to the digits, and the tendons of the flexor muscles, are delineated.

In making the dissection the integuments and the deep palmar fascia are first to be removed. The former may be raised by a cut along the centre of the palm, terminated by cross cuts at the wrist and the roots of the fingers; and as the inner flap is raised, the palmaris brevis muscle will be met with in the fat. After the palmar fascia has been denuded, and its arrangement at the fingers examined, it may be cut behind, where it joins the tendon of the palmaris longus, E, and may be thrown forwards.

By taking away the teguments of one finger, say the middle, the sheaths of the flexor tendons will come into view; and after the removal of the sheath, the arrangement of the tendons will be manifest, as in the ring finger.

CENTRAL MUSCLES OF THE PALM.

In the hollow of the hand lie the flexor tendons, with some other muscles. Laterally the muscles of the thumb and little finger form on each side a ball or prominence, to be noticed afterwards; and the group on the inner side is partly covered by the small subcutaneous palmar muscle.

A. Palmaris brevis.
B. Abductor pollicis.
D. Flexor brevis pollicis (outer head).
E. Tendon of palmaris longus.
G. Adductor minimi digiti.
H. Adductor pollicis.
J. Abductor minimi digiti.
K. K. Pieces of the sheath of the flexor tendons.
L. Part of the palmar fascia.

N. First dorsal interosseous muscle.
O. O. Two outer lumbricales.
R. Tendons of the flexor digitorum sublimis.
S. Tendon of flexor carpi ulnaris.
V. Flexor sublimis tendons in the palm of the hand.
W. Tendon of flexor profundus to the ring finger.
X. Anterior annular ligament.

Palmaris brevis, A. This small subcutaneous muscle is unattached to bone. Consisting of fleshy bundles, more or less separate, which are attached to the palmar fascia, L, it is inserted into the skin at the inner border of the hand, extending downwards a varying distance from the pisiform bone. Its insertion is marked by a surface depression.

When the muscle contracts it elevates the skin on the inner side of the hand, and increases slightly the depth of the palmar hollow.

The tendon of the *palmaris longus*, E, enters the hand over the annular ligament: from its outer side an offset is prolonged to the thumb muscles, whilst the main part ends in the palmar fascia.

Tendons of the flexor sublimis digitorum, V. Four in number, they are directed through the palm over the deep flexor; and at the root of each finger one enters the sheath of the digit, K, with a tendon of the deep flexor. Near the front of the metacarpal phalanx it is slit for the passage of the deep flexor tendon, W; and it is inserted by two slips into the sides of the second phalanx, about half way along the bone.

This muscle brings the middle phalanges towards the palm, and bends thus the nearest phalangeal joints—the first stage in the movement of closing the fingers. As the fingers are approximated to the palm, the muscle raises the metacarpal phalanges by means of the digital sheaths binding its tendons to the bones; and it acts finally as a flexor of the wrist-joint.

Tendons of the flexor profundus (Plate ix.), also four in number, cross the palm beneath the superficial flexor, and may be seen projecting slightly on the sides. Entering the digital sheaths, each is transmitted through the accompanying flexor sublimis tendon, as is shown on the ring finger, and passes onward to be inserted by a single piece into the base of the last phalanx. Small rounded muscles, the lumbricales, are attached to these tendons in the palm (Fig. ii.).

Between each tendon of the deep flexor and the fore part of the middle phalanx is a thin membranous band (opposite W) uniting the two, which is called "ligamentum breve;" and intervening in like manner between each piece of the superficial flexor and the front of the metacarpal phalanx, is another "ligamentum breve," to fix this tendon to the underlying bone.

The deep flexor draws forward the last phalanges, and bends the last phalangeal joints. Still continuing to shorten, it assists the superficial flexor in bending the metacarpo-phalangeal joints in the act of shutting the fingers; and combined with the same muscle, it will flex the wrist when the digits are closed.

In amputating on the living body through the phalangeal articulations, some difficulty is experienced, when the joint is opened at the back, in entering the knife between the ends of the bones, owing to the flexor tendon drawing the distal against the nearer phalanx; and this difficulty is increased in the case of the last joint, in consequence of the smallness of the part to be held preventing sufficient force being employed to overcome the tendon. When the joint is opened at the front the impeding tendon has been previously cut, and the operation can be executed without hindrance to the passage of the knife.

Sheath of the flexor tendons, K. In each finger this reaches from the palm of the hand to the last phalanx. It is constructed on the one side by the bones; and on the other by fibrous bands, which are thinnest opposite the joints: these thinner pieces have been removed in the dissection.

A synovial membrane lines each sheath, projecting into the palm of the hand, where it is closed; and long tapering folds (vincula vasculosa) are continued from it to the tendons: one of these, connected with the deep flexor, is shown in the opened sheath of the ring finger. In the thumb and the little finger the synovial membrane of the sheath is continued upwards into a large synovial sac which surrounds the tendons of both flexors beneath the annular ligament.

SUPERFICIAL ARTERIES OF THE HAND.

The arrangement of the superficial palmar arch and its offsets, which is described as the usual one, is figured here, but many hands were examined before this condition was found. The arteries to the thumb and the radial side of the fore finger will be described in the explanation of Fig. ii.

a. Ulnar artery in the forearm.
b. Radial artery in the forearm.
c. Superficial palmar arch.
d. Superficial volar branch.
f. Four digital branches of the superficial palmar arch.

g. Communicating artery from the palmar arch to the radial branch of the index finger.
h. Communicating branch to the deep arch from the digital artery of the little finger.

The *ulnar artery*, *a*, enters the hand over the annular ligament, and curving towards the ball of the thumb forms the superficial palmar arch: it supplies branches to the greater number of the digits.

The *superficial palmar arch*—the continuation of the ulnar artery—lies across the hollow of the hand, between the lower border of the annular ligament and the ball of the thumb. With its convexity towards the fingers, it reaches nearly as far forwards as a line, across the palm, from the middle of the fold between the thumb and the forefinger. Diminishing in size, it ends externally by joining branches of the radial, viz., the superficial volar branch, *d*, pretty constantly, and the branch to the radial side of the forefinger (Fig. ii., *d*) occasionally, by means of the small communicating branch, *g*.

At its inner end the arch is covered by the palmaris brevis muscle, A, and thence to the ball of the thumb, by the integuments and the palmar fascia; it rests on the tendons of the flexors of the digits, and on the ulnar and median nerves. Companion veins lie on the sides of the artery.

From the concavity of the arch spring small unnamed offsets; and from the convexity digital arteries proceed.

The *digital arteries,* four in number, and marked each with the letter *f,* supply three digits and a half. In their course to the digits the three outer lie over interosseous spaces, whilst the other is placed along the inner part of the palm; and at the cleft of the fingers they divide, except the most internal, into two for the contiguous sides of the digits. Coursing along the fingers they are united by a loop behind each phalangeal joint; and at the end of the finger they terminate in a loop which gives offsets to the tip, as is seen on the fourth digit.

The following communications take place between the digital arteries of the ulnar and the branches of the radial. At the inner side of the palm the branch *h,* which springs from the artery to the inner side of the little finger, joins either the deep arch or an interosseous branch; at the roots of the fingers the digital arteries anastomose with the interosseous branches of the deep arch; and at the tip of the forefinger the digital artery on the ulnar side inosculates with the arteria radialis indicis.

In the hand the large digital vessels and nerves lie over the intervals between the metacarpal bones; and in the fingers they occupy the sides. Incisions into the palm of the hand can be made therefore with least injury over the line of the metacarpal bones; and a cut into a finger, along its centre.

Wounds of arteries in the palm of the hand are followed generally by copious bleeding, in consequence of the numerous communications between the chief vessels. In an injury of the superficial palmar arch, at *c,* for instance, blood will be furnished by the ulnar trunk, *a.* And though this source might be cut off by a ligature, the blood could be supplied by the radial artery to the other end of the arch, through the anastomosing branches, *d* and *g;* or through the anastomoses above described of the digital with the interosseous arteries, and with the arteria radialis indicis. In such an arrangement of the vessels as that delineated in the Figure, the bleeding from the wound might be commanded by placing a ligature on each side of the orifice in the artery; or, as is more commonly done, by stopping the currents in the radial and ulnar trunks by pressure above the wrist, and by applying a graduated compress to the seat of injury. If, when the orifice of the artery has not been secured by a thread, pressure has been found ineffectual in stopping the bleeding, ligature of the ulnar artery, or of this and the radial, would have to be performed in addition to a compress to the wound.

But there is an occasional condition of the vessels, which renders the arrest of the hæmorrhage difficult unless the artery is tied in the wound. For instance, a third artery, sometimes as large as either the radial or the ulnar, may join the middle or the outer part of the superficial palmar arch, so as to bring blood freely to the wound. And as this vessel (usually an offset of the anterior interosseous, but sometimes of the brachial or the radial*) courses with the median nerve beneath the annular ligament, and generally beneath the muscles, pressure would not be productive of much benefit in stopping the current in it, and ligature of it would be scarcely practicable. Recurring bleeding with the existence of this state of the vessels would be quite uncontrollable by means which would arrest it when the ordinary arrangement existed.

As the state of the palmar wound is sometimes unfavorable to any attempt to place a ligature on the vessel there, and as surgeons have a reasonable disinclination to enlarge wounds of the palm to search for the bleeding orifice, ligature of the brachial trunk has been practised when the bleeding resists the usual means of stopping it. The following case, illustrative of the inefficacy of securing the main trunk of the limb at a distance from the wound when a large branch joins the arch directly, as in the condition stated above, is instructive.

"A young man wounded his palmar arch: secondary hæmorrhage took place several times; the radial and ulnar were tied, but the bleeding returned; an artery of some size, a 'vas aberrans,' was discovered beating in the middle of the forearm, close under the skin; a ligature was put on the brachial in the middle of the arm, with the hope of getting above the origin of the abnormal branch; it (the unusual branch) continued however to pulsate after the ligature was tightened; the vas aberrans itself was therefore tied at once, close below the elbow, but notwithstanding all these precautions, hæmorrhage occurred on the following day as violent as ever; the wound (in the palm) was a second time enlarged, and fortunately the blood burst forth at the time of operation and the wounded artery was easily tied: the patient recovered rapidly."†

---

* Examples of these conditions of the arteries, collected chiefly by Mr. Quain, are contained in the museum of University College, London.

† This case is reported by Mr. Cadge, in the part of Morton's Surgical Anatomy

The result of ligature of the brachial in the above-cited case teaches, that tying the vessel in the wound of the palm, if such a step is possible, is to be preferred to a distant operation on the main artery of the limb, in those instances in which the surgeon suspects that a large unusual median artery joins the superficial palmar arch.

SUPERFICIAL NERVES OF THE HAND.

The median and ulnar nerves divide in the palm of the hand into large branches, which end on the digits as the nerves of touch. They give branches to the superficial muscles; and the ulnar nerve supplies also the deep muscles by means of a special offset.

1. Trunk of the median nerve.
2. First digital branch.
3. Second digital branch.
4. Third digital branch.
5. Fourth digital branch.
6. Fifth digital branch.
7. Communicating branch from the median to the ulnar.
8. Outer digital branch of the ulnar.
9. Inner digital branch of the ulnar.
10. Trunk of the ulnar nerve.

The *median nerve*, 1, is the larger of the two trunks distributed in the hand. Issuing from beneath the annular ligament, it is consumed chiefly in five digital branches which supply both sides of each of the three outer digits, and the outer side or half of the ring finger. Comparatively few branches are furnished to muscles.

The *digital branches* are continued through the palm of the hand, and along the sides of the digits to the extremity, where they end in a tuft of offsets for the supply of the ball and nail-pulp of the finger. To the skin of the palm and the surface of the digits they give many branches.

Two of them supply lumbrical muscles; thus the third nerve, 4, gives a branch to the most external lumbricalis; and the fourth nerve, 5, to the next following muscle.

Muscular branches. Part of the fleshy ball of the thumb is supplied

---

which was completed by him (London, 1850, p. 371). The artery named "vas aberrans," does not correspond with the arteries commonly so called; and it was probably the "median artery," which sometimes arises from the lower end of the brachial, and joins the palmar arch, as Plate 43 of Mr. Quain's Work on the Surgical Anatomy of the Arteries illustrates.

by the branch, 8; this is distributed to the muscles outside the long flexor tendon, viz., to the abductor pollicis, B, opponens pollicis, C, and the outer head of the short flexor, D.

If the median nerve was cut through close to the annular ligament, sensibility would be destroyed in the palmar surface of the hand and fingers outside a line drawn from the middle of the ring finger to the wrist; and it would be diminished at the dorsal aspect of the three outer digits beyond the matacarpo-phalangeal joints, where offsets from the digital nerves ramify.

The muscles of the thumb before referred to as supplied by the median, and marked B, C, and D, together with the outer two lumbricales, would be paralyzed.

The *ulnar nerve*, 10, divides on the annular ligaments into a superficial or digital, and a deep or muscular part.

From the superficial part two *digital branches*, 8 and 9, are furnished to the little finger (both sides), and to half the ring finger; these have a similar distribution to the digital nerves of the median.

The branch marked 9 sends offsets to the palmaris brevis muscle, and the integuments of the inner part of the hand; and the external of the two, 8, receives a connecting branch, 7, from the median nerve.

Insensibility of the palmar surface of the hand and fingers, inside a line from the ring-finger to the wrist, follows incision of the trunk of the ulnar nerve; and the power of feeling would be lost at the same time on the back of the two inner fingers which are supplied by the same nerve. Besides the paralysis of the deep muscles attending injury of the nerve, which will be noticed in the description of Fig. ii., the small palmaris muscle, A, will lose its power of contracting.

Figure II.

Most of the special muscles of the hand, and the deep palmar arch, with its companion nerve, are represented in the right-hand Figure of the Plate.

This dissection follows the preceding; and to carry it out, the superficial palmar arch and the ulnar and median nerves are to be cut through at the annular ligament, and are to be taken away: then the superficial and deep flexor tendons having been cut at the same spot, are to be thrown forwards to the digits—the lumbrical muscles attached to the deep tendons being cleaned as the superficial tendons are raised.

SPECIAL MUSCLES OF THE HAND.

All the muscles which have both origin and insertion in the hand will be now described, with the exception of the palmaris brevis. They consist of three sets: a thumb group, a little-finger group, and a central group for the other digits.

- B. Abductor pollicis.
- C. Opponens pollicis.
- D. Outer head of flexor brevis pollicis.
- F. Inner head of flexor brevis.
- H. Adductor pollicis.
- J. Abductor minimi digiti.
- K. Flexor brevis minimi digiti.
- L. Adductor minimi digiti.
- M. Tendon of the flexor longus pollicis.
- N. First dorsal interosseous.
- O. Lumbricales muscles.
- P. Interossei of the hand.
- R. Tendon of flexor digitorum sublimis.
- S. Tendon of flexor carpi ulnaris.
- T. Tendon of flexor carpi radialis.
- V. Tendons of flexor profundus.

The *thumb muscles*, four in number, consist of an abductor, an adductor, and a flexor, with a special muscle to oppose the thumb to the other digits.

Abductor pollicis, B. This is the most superficial muscle. It arises behind from the annular ligament, and the ridge of the trapezium bone; and is inserted by a tendon into the outside of the base of the first phalanx.

The muscle draws away the thumb from the index finger.

Adductor pollicis, H. The origin of the muscle, which is not always separate from the inner head of the flexor brevis, is fixed to the ridge on the palmar surface of the metacarpal bone of the middle finger; and the muscle is inserted with the inner head, F, of the short flexor into the inner side of the base of the first phalanx.

By its action the thumb is placed on the palm and the fore finger, so as to deepen externally the hollow of the hand.

Opponens pollicis, C. The muscle is partly covered by the abductor, and arises, like it, from the annular ligament and the prominence of the trapezium bone: it is inserted into the metacarpal bone of the thumb along the outer edge.

This muscle can abduct the metacarpal bone, and can then so move it as to allow the ball of the thumb to be turned opposite the ball of each digit, as in the act of picking up a pea with the thumb and each finger.

The *flexor brevis pollicis* is divided into two pieces or heads, D and F, at its insertion into the thumb. Single at its hinder attachment or origin, it is fixed to the annular ligament near the lower edge, to two carpal bones (os trapezoides and os magnum), and to the bases of the two metacarpal bones answering to the two carpals. The fibres, collected into two bundles which are separated by the tendon, M, of the long flexor, are inserted into the sesamoid bones, and the base of the first phalanx— the outer head, D, joining the abductor, and the inner head, F, blending with the adductor pollicis.

This muscle bends the metacarpo-phalangeal joint of the thumb; it draws inwards also the thumb over the palm and approaches it to the other digits.

Little-finger muscles. The group of muscles connected with the little finger contains three, viz., an abductor, an adductor, and a short flexor, as in the thumb; but the flexor is sometimes absent.

Abductor minimi digiti, J, arises behind from the pisiform bone ; and is inserted into the base of the first phalanx on the inside, sending an offset to join the extensor tendon.[*]

It draws the little from the ring finger, and assists in bending the metacarpo-phalangeal joint.

Adductor minimi digiti, vel opponens, L, arises posteriorly from the annular ligament, and the hook of the unciform bone; and it is inserted into the inner side of the fifth metacarpal bone.

The fibres shortening as they contract, draw forwards the metacarpal bone, and deepen the hollow of the palm.

Flexor brevis minimi digiti, K, takes origin from the annular ligament and the unciform process, superficial to the preceding muscle; it is inserted with the abductor into the base of the first phalanx.

By its position in the hand this muscle is enabled to act as a flexor of the metacarpo-phalangeal joint.

In the *central group* of the hand are included superficial and deep muscles: the former, or the lumbricales, are attached to the deep flexor

---

[*] Lehre von den Muskeln, etc. Von Friedrich Wilhelm Theile. Leipzig, 1841, p. 283.

tendons; and the latter, the interossei, lie between the metacarpal bones.

The *lumbrical muscles,* four in number, and marked by the letter O, arise from the tendons of the flexor digitorum profundus, near the wrist. Placed on the radial side of the flexor tendons, each joins, opposite the metacarpo-phalangeal joint, the extensor tendon on the back of the first phalanx. The two external muscles arise each from a single tendon, and the two internal from two tendons for each.

They flex the metacarpo-phalangeal joints by bringing forwards the first phalanges, and assist the special flexors in closing the fingers.

The *interossei muscles* occupy the inter-metacarpal spaces—two being present in each space except the first, in which there is only one: they are divided into two sets, palmar and dorsal.

The *dorsal set,* four in number, are shown in Plate xi. Each arises from the two bones bounding the metacarpal space, and is inserted into the base of the metacarpal phalanx, chiefly into the bone, though it joins also the extensor tendon by a fibrous process. The first or most external, which is sometimes called abductor indicis, is the largest: it is marked by N.

These muscles act as abductors of the fore and ring fingers from the middle one; and they can move the last-mentioned digit to each side of a line passing lengthwise through it. The first may adduct the metacarpal bone of the thumb to that of the index finger.

The *palmar set,* only three in number, lie in the three inner spaces; and the middle finger does not receive any of this set. Arising from the metacarpal bone of the finger to which each belongs, they are inserted, like the dorsal, into the nearest phalanx of the fingers; each has but a slight attachment to the bone, blending most with the extensor tendon.*

When acting they bring together the separated fingers, and will draw the fingers, into which they are inserted, viz., the fore, ring, and little, towards the middle digit.

DEEP ARTERIES OF THE HAND.

The radial artery ends in the palm of the hand by forming an arch; and it furnishes arteries to the digit and a half left unsupplied by the

---

* This difference in the insertion of the dorsal and palmar sets, is stated by Theile in the work on the muscles before quoted, p. 286.

ulnar trunk. It enters likewise into numerus communications with offsets of the ulnar artery.

- a. Deep palmar arch.
- b. Profunda branch of the ulnar artery.
- c. Large artery of the thumb.
- d. Digital artery to the radial side of the fore finger.
- e. Communicating branch of the deep arch from the digital artery to the little finger.
- f, f. Two inner interosseous arteries.
- h, h. Offsets to lumbrical muscles.

Radial artery. Passing into the hand through the first interosseous space, it furnishes digital arteries to the thumb and the fore finger, and ends in the deep palmar arch.

The *digital artery of the thumb*, c (art. magna pollicis), courses along the metacarpal bone, and divides into two branches near the joint between that bone and the phalanx: these supply the sides of the thumb, and join at the tip in the usual way.

The *digital artery of the index finger*, d (art. radialis indicis), lies along the second metacarpal bone; and issuing from beneath the adductor pollicis, H, runs on the radial border of its digit to the extremity, where it anastomoses with the digital artery from the superficial palmar arch. Sometimes it joins the superficial palmar arch through a branch, g, Fig. I.

The *deep palmar arch*, a, is the curve formed by the end of the radial artery. It reaches from the first to the fourth interosseous space, but placed near the carpus, and is rather convex forwards, like the superficial arch. At the inner end it communicates with the profunda branch, b, of the ulnar; and with the branch, e, belonging to the digital artery of the inner side of the little finger. The arch has a deep position in the hand;—internally it is covered by the adductor minimi digiti, L; externally by the inner head of the flexor brevis pollicis, F; and between these, by the flexor tendons: it rests on the three middle metacarpal bones, and their intervening muscles.

Offsets of the arch.—From the concavity small offsets are directed back to the carpus. Three small *perforating arteries* pierce the three inner dorsal interosseous muscles to reach the back of the hand. The chief branches are described below:—

Interosseous branches, f, f.—Only two of these now appear, and the third lies beneath the adductor pollicis: they extend to the clefts of the fingers, giving muscular twigs, and end by joining the digital arteries of the superficial arch.

Muscular branches, h, h, supply the two or three inner lumbrical muscles.

Through the communication of the radial with the ulnar artery at the inner side of the hand, blood would find its way directly into the superficial palmar arch, after ligature of the ulnar trunk above the wrist; and by means of the anastomoses between the branches of the two arteries (p. 82), the blood would be conveyed from the superficial to the deep arch.

Wounds of the deep arch are rare, in consequence of its deeper and securer position in the hand; and ligature of the vessel would be oftentimes impossible. Supposing the wounded vessel cannot be reached, the bleeding would most commonly be arrested by a graduated compress to the wound, and by pressure on the radial and ulnar arteries in the lower third of the forearm. If those means fail to stop the bursting forth of the blood, ligature of the radial artery—the chief vessel entering the arch—would probably be effectual in commanding the hæmorrhage. Should bleeding still occur, and possibly from large communicating branches with the ulnar artery, for no large unusual artery joins the deep arch, tying the ulnar trunk might be tried. As a last resource ligature of the brachial artery remains.

DEEP NERVE OF THE PALM.

The ulnar nerve is distributed to those muscles of the inner and deep parts of the palm of the hand, which do not receive branches from the median nerve.

1, 1. Branches to the lumbrical muscles.	2. Deep palmar branch of the ulnar nerve.
	3. Trunk of the ulnar nerve.

The *deep palmar branch*, 2, of the ulnar nerve, arising near the wrist, passes deeply between the flexor brevis and abductor minimi digiti, or through the adductor muscle, L, as in the Figure, and accompanies the radial arch to the first interosseous space, where it ends by supplying the adductor pollicis, H, and the inner head, F, of the flexor brevis pollicis.

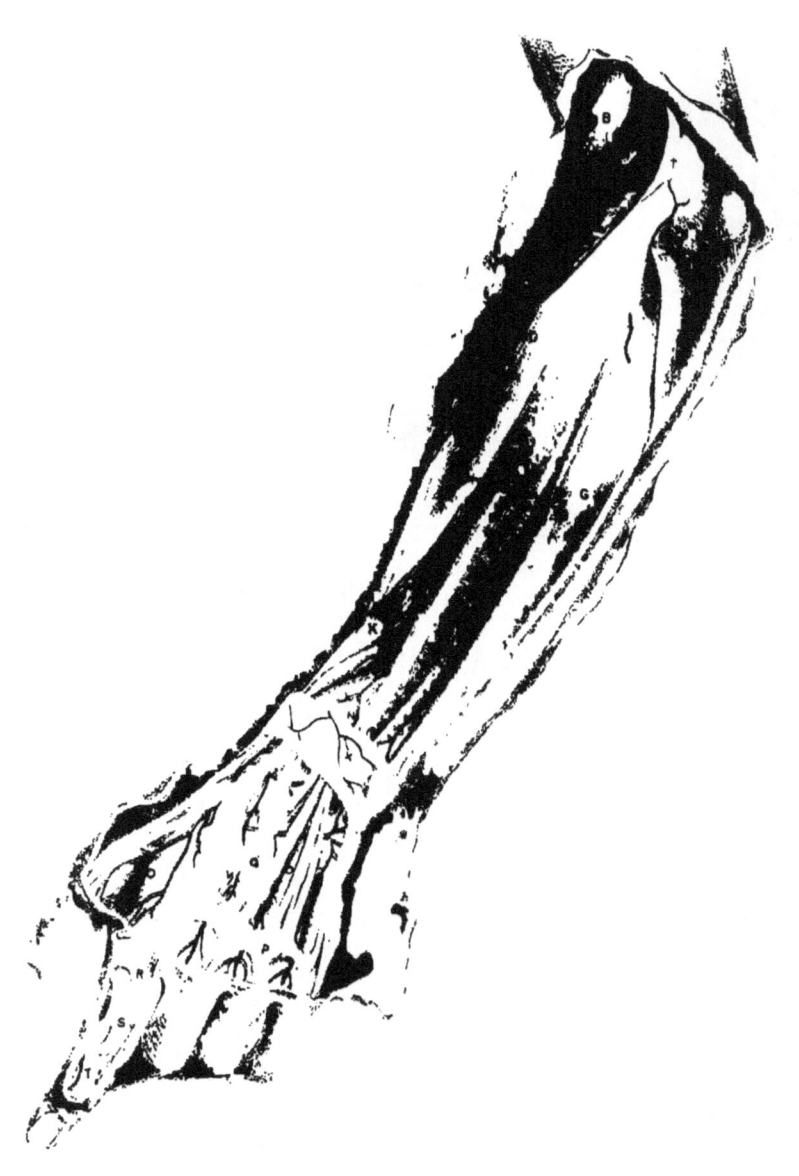

Muscular offsets are furnished to the muscles of the little finger, viz., adductor, J, flexor brevis, K, and adductor, L; to all the seven interossei muscles; and to the inner two lumbricales.

All the muscles of the hand, except two and a half of the thumb and the two outer lumbricales, receive branches from the deep part of the ulnar nerve. Destruction of the trunk of the ulnar nerve in the arm would affect the movements of the thumb and fingers; but notably those of the little and ring fingers, whose short or hand muscles depend solely on the ulnar nerve for their contractile power.

DESCRIPTION OF PLATE XI.

The dissection of the superficial muscles and vessels on the back of the forearm and hand is here illustrated.

This view was obtained by reflecting the integuments from the elbow to the roots of the fingers; and by removing the deep fascia, with the exception of the posterior annular ligament near the wrist. The fore finger was then denuded of its cutaneous coverings, to trace the extensor tendon to the end.

SUPERFICIAL MUSCLES.

At the back of the forearm are located the muscles which oppose by their action the muscles in front; and as the anterior group consists of flexors and pronators, so the posterior includes their antagonists—extensors and supinators.

The posterior set is divided, like the anterior, into superficial and deep layers. In the superficial layer are contained one supinator, and the extensors of the wrist and digits, which are indicated below by the letters of reference.

A few of the deeper muscles appear near the wrist, but these will be described with Plate xii.

A. Biceps flexor brachii.
B. Supinator longus.
C. Extensor carpi radialis longior.
D. Extensor carpi radialis brevior.
E. Extensor digitorum communis.
F. Extensor minimi digiti.
G. Extensor carpi ulnaris.
H. Anconeus.
J. Brachialis anticus.
K. Extensor ossis metacarpi pollicis.
L. Extensor primi internodii pollicis.
N. Extensor secundi internodii pollicis.
O. Dorsal interosseous muscle.
P. Fibrous bands joining the extensor tendons near the knuckles.
R. Expansion from the extensor tendon opposite the finger joints.
S. Splitting of the extensor tendon.
T. Insertion of the extensor tendon into the last phalanx.
V. Tendon of the indicator muscle.
X. Posterior annular ligament.
† Deep fascia of the arm.

The *supinator longus*, B ("brachio-radialis," Sœmmer.), is the most external muscle; it appears also in the anterior view of the forearm, with the description of which (Plate viii.) part of its anatomy has been given.

Arising, as before said, from the upper two thirds of the condyloid ridge of the humerus, and from the intermuscular septum, it is inserted into the radius close to the root of the styloid process.

Narrowed at the origin, it is widened below the elbow over the subjacent muscles forming the prominence on the outer side of the forearm. The anterior border touches the brachialis anticus, J, the biceps, A, and the pronator teres (Plate viii.); and the posterior edge is in contact with the extensor carpi radialis longior, C, and with the extensor carpi radialis brevior,* D. Near its insertion the tendon is covered by the extensors of the thumb.

This supinator acts mostly as a flexor of the elbow joint. If the hand

---

* This projection backwards of the supinator so as to touch the extensor carpi radialis brevior is not referred to by anatomists of authority. It is not represented by Albinus in his standard work, Tabulæ Anatomicæ Musculorum Hominis. Lond., 1747; nor in the modern work of Bourgery and Jacob, Traité complet de l'Anatomie de l'Homme. Paris, 1833. Neither Theile, in his treatise on the muscles in Sœmmerring's Anatomy (Lehre von den Muskeln, etc. Leipzig, 1841), nor Henle, in his recent Handbuch der Systematischen Anatomie des Menschen; Dritte Abtheilung, Braunschweig, 1858, takes notice of the fact. Cruveilhier is silent also respecting this connection of the muscle in his systematic work, Traité d'Anatomie descriptive. Deuxième édition. Paris, 1843.

is greatly pronated, the muscle can draw backwards the radius to a small extent; and if the hand is much supinated, the lower end of the radius will be moved somewhat forwards as in pronation: in both cases the hand is brought into a state midway between pronation and supination (Theile).

The *extensor carpi radialis longior*, C, arises from the lower third of the outer condyloid ridge of the humerus, and, below the elbow, from the intermuscular septum between it and the following extensor. In the lower part of the forearm its tendon passes through the posterior annular ligament with the shorter extensor, and is inserted into the base of the metacarpal bone of the index finger.

The muscle is superficial above and below; but it is covered by the supinator longus in the upper part of the forearm.

The *extensor carpi radialis brevior*, D, takes origin from the outer condyle of the humerus by the common tendon,* and from the capsule of the elbow joint. Beyond the annular ligament the tendon is inserted into the base of the metacarpal bone of the second finger.

This extensor is superficial in great part, but two muscles of the thumb, K and L, rest on it below. Along the outer edge lie the long radial extensor of the wrist, and the long supinator.

Both radial extensors draw backwards the hand, extending thus the wrist. The longer muscle can assist the supinator in bending the elbow; and the shorter one may help in straightening the elbow after the joint has been bent.

Extensor digitorum communis, E. Attached above by the common origin, it ends below in four tendons: these cross the back of the hand, and are inserted into the middle and ungual phalanges of the fingers.

On the hand the little finger tendon is often united in part with that of the ring finger. Near the knuckles all are joined by lateral bands; but those of the ring-finger tendon being stronger than the rest, prevent extension of that digit whilst the fingers on the sides (little and middle) are bent.

On each finger the tendon forms a common expansion over the first

---

* This common tendon is fixed to the lower part of the condyle and sends downwards aponeurotic septa on the under and lateral surfaces of three other muscles, viz., the extensor digitorum communis, extensor minimi digiti, and extensor carpi ulnaris.

phalanx with the tendons of the lumbricales and interossei.* At the front of the phalanx this expansion divides into three,† S: of these, the central part is fixed into the second phalanx at the base: while the two lateral pieces join, and are inserted as one, T, into the base of the last phalanx. Opposite each phalangeal joint a fibrous expansion is continued from the tendon to the capsule· on the first joint this is indicated by the letter R.

When straightening the fingers the muscle extends the joints from root to tip, separating the digits at the same time; it acts secondarily as an extensor of the wrist. If the elbow has been bent, it can become an extensor, like the other muscles which take origin by the common tendon.

The *extensor minimi digiti*, F, more or less united with the preceding, is sometimes tendinous in the upper third of the forearm, as is shown in the Plate. Arising by the common attachment, its tendon is divided into two beyond the annular ligament, and the pieces blend on the first phalanx with the other tendons.

The muscle extends the little finger, and exercises afterwards the same action on the wrist joint.

Extensor carpi ulnaris, G. With the common origin above, the muscle is fixed also by aponeurosis to the ulna for three inches below the anconeus, H. Passing through the annular ligament, it has a tendinous insertion into the base of the metacarpal bone of the little finger.

The hand is drawn backwards and to the ulnar side by this muscle.

Anconeus, H. This, the smallest of the superficial muscles, arises from the hinder and lower parts of the condyle of the humerus, and chiefly by a separate tendon. The fibres give rise to a belly of a triangular shape as they are directed downwards and inwards to their insertion into the upper third of the ulna, on the posterior surface. Some of the upper fleshy fibres seem continuous with the fibres of the triceps.

Inserted into the ulna it will draw backwards this bone—the humerus being fixed—and will extend the elbow-joint in conjunction with the triceps.

Extensors of the thumb. Three muscles, extending the thumb, issue

---

* On the back of the fore and ring-fingers the special extensors of those digits blend with the common expansion.

† In the natural state a thin membrane connects the pieces of the tendon, but this was removed in the dissection to render more evident the arrangement above described.

between the common extensor of the fingers, E, and the radial extensors of the wrist, C and D. Two, viz., the extensor of the metacarpal bone, K, and that of the first phalanx, L, lie close together on the outer border of the forearm: and the third, the extensor of the last phalanx, N, is placed below the others, and is separated from them by an interval. The anatomy of these muscles will be found in the description of Plate xii.

Indicator muscle, V. Only below the annular ligament is the tendon of this muscle visible; and it blends with the common expansion on the first phalanx of the fore finger.

The *posterior annular ligament*, X, confines the tendons of the muscles to the wrist, so as to make the extensors of the digitis carry backwards the hand after the digits have been straightened. Formed mostly of transverse fibres, but continuous above and below with the special fascia of the limb, it is fixed externally into the radius and internally, where it reaches lower down, into two bones of the carpus—cuneiform and pisiform.

As the tendons pass beneath this band they are lodged in separate channels. There are six spaces, in which the tendons are arranged in the following order:—The most internal compartment contains the extensor carpi ulnaris, G; and the one that follows on the outer side is occupied by the extensor minimi digiti, F. The next space receives the common extensor of the fingers, E, and the special extensor of the fore finger, V; and then comes a narrow sheath for the extensor secundi internodii pollicis, N. Still to the radial side is a large space lodging the two radial extensors of the wrist, C and D; and most external of all is the tube through which pass the extensor ossis metacarpi pollicis, K, and extensor primi internodii pollicis, L. Each sheath in the ligament is provided with a synovial membrane.

All the tendons, with one exception, lie in grooves in the subjacent bones, and to the edges of the grooves processes of the fibrous tissue are attached. The tendon not resting on the bone is that of the extensor minimi digiti, F, which lies between the radius and ulna. On the radial side of that extensor the tendons groove the radius in the order stated, and on the ulnar side, one muscle (ext. carpi ulnaris) is lodged in a hollow on the ulna.

The *dorsal interosseous muscle*, O, arising from the metacarpal bones bounding each space, are pierced behind by vessels—the external one by the radial trunk, and the others, by the perforating branches from the deep palmar arch. The attachments and the action of these muscles are described with Plate X.

ARTERIES OF THE BACK OF THE HAND.

About the wrist, and on the back of the hand, the arteries are derived from the radial and interosseous vessels, and from the deep palmar arch. Above the wrist only superficial branches of the interosseous vessels appear.

a. Radial artery.
b. Posterior carpal branch.
c. Metacarpal branch.
d. Dorsal branch of the thumb and the index finger.
f. Dorsal interosseous arteries.
g. Branch of the posterior interosseous artery.

h. Posterior part of the anterior interosseous artery.
k. Offset of the recurrent interosseous artery.
† † † Cutaneous offsets of the posterior interosseous artery.

The *radial artery, a*, corresponding with the dorsal artery of the foot in the lower limb, winds over the carpal bones and enters the hand through the first interosseous space. Its connections are the following:—

In addition to the common investments of the limb, with superficial veins and nerves, the three extensors of the thumb are directed over it;—two, viz., extensor of the metacarpal bone, K, and of the first phalanx, L, lie close together, and in a line with the styloid process of the radius; and the other, the extensor of the second phalanx, N, crosses close to the spot where it enters the palm of the hand. Beneath the artery are the carpus and the external lateral ligament of the wrist-joint.

Small veins, and ramifications of the external cutaneous nerve, accompany the artery.

Its branches are inconsiderable in size, but numerous, and are distributed to the back of the hand and some digits.

The place of the radial artery can be easily ascertained through the skin, if the tendons crossing it are made prominent by extension of the thumb; and as the vessel is closer to the extensor of the second phalanx than to the others, this tendon should be taken as the guide to it.

Slight wounds on the back of the wrist would be likely to open the artery; and when the radial lies over the tendons instead of under them, it is still more superficial, and is more exposed to accident.

In disarticulation of the metacarpal bone of the thumb, the artery lies close to the joint, and will be cut unless the knife is kept near the bone.

Branches of the artery supply the carpus, the metacarpus, and the digits.

The *posterior carpal branch, b,* forms an arch behind the wrist with a corresponding branch of the ulnar artery, and communicates with the posterior interosseous, *g:* from this carpal arch interosseous arteries are sometimes given to the inner two metacarpal spaces.

The *metacarpal branch, c,* arising here in common with the preceding, runs to the second interosseous space, and ends at the front of the space in two branches for the contiguous sides of the fore and middle digits on the dorsal surface. Behind it receives a perforating branch from the deep palmar arch, and in front it communicates with the digital arteries.

Dorsal interosseous arteries, f, f, lie over the inner two interosseous muscles, and are derived from the dorsal carpal arch; or they may come from the perforating arteries of the deep palmar arch, as in the dissection from which the Drawing was made. At the cleft of the fingers they give offsets to the sides of the digits, and anastomose with the digital arteries; and if they spring from the dorsal carpal arch, they receive, behind, the perforating arteries from the deep palmar arch.

Dorsal branches of the thumb and fore finger.—Two small branches belong to the thumb, and these run along the metacarpal bone—one on each side, to the last phalanx: the inner one of these is marked *d;* and the outer one springs from the radial trunk, about half an inch higher up. There is one branch for the fore finger, which is continued on the radial side of that digit, and supplies the integuments; in this body it is conjoined with the inner artery to the dorsum of the thumb.

Both the posterior interosseous artery, *g,* and the anterior interosseous, *h,* appear near the wrist; but they belong to the deeper dissection, with which they will be described.

DESCRIPTION OF PLATE XII.

The deep muscles of the back of the forearm and the posterior interosseous artery and nerve are pictured in this Plate.

The superficial muscles have been cut through near their origin, with the exception of the supinator longus on the one side, and the anconeus on the other. In reflecting the extensors of the fingers, the branches of vessels and nerves to them should be defined at the same time.

DEEP MUSCLES.

In the group of deep muscles at the back of the forearm are included three extensors of the thumb, the special extensor of the fore finger, and the short supinator.

A. Supinator brevis.
B. Extensor ossis metacarpi pollicis.
C. Extensor primi internodii pollicis.
D. Extensor secundi internodii pollicis.
E. Extensor proprius indicis.
F. Extensor longus digitorum, cut.
G. Extensor carpi ulnaris.

H. Anconeus muscle.
K. Extensor carpi radialis brevior.
L. Extensor carpi radialis longior.
M. Supinator longus.
N. Brachialis anticus.
P. Biceps brachii muscle.
R. Triceps brachii muscle.
S. Posterior annular ligament.
† External lateral ligament of the elbow-joint.

The *supinator brevis*, A, nearly encircles the upper part of the radius, and is the highest of the deep muscles. It arises from the ulna below the small sigmoid notch, from the orbicular ligament of the radius, and from the external lateral ligament of the elbow-joint. The fibres curve forwards and downwards, and are inserted into the radius so as to cover that bone as low as the pronator teres, except along a triangular surface on the inner side: the lowest fibres taper to a point externally, and the highest inclose the neck of the radius.

The connections of the supinator with muscles, vessels, and nerves, are numerous and complicated. An anterior view of the muscle is given in Plate ix. Perforating the muscular fibres is the posterior interosseous nerve; and the posterior interosseous artery appears at the lower border.

The muscle turns the upper end of the radius backwards, and supinates the hand. It is the direct antagonist with the biceps of the pronator teres; and in consequence of the attachment of both muscles near the upper end of the radius, they keep the upper fragment supinated in fracture of the shaft of the bone.

Extensor ossis metacarpi pollicis, B, the largest of the thumb extensors, arises from both bones of the forearm, and from the interosseous membrane, viz., from three inches of the radius below the supinator, and from a narrowed surface of the ulna of about the same length, and close to the outer edge.

In company with the next extensor it occupies the outer compartment of the annular ligament; and it is inserted into the base of the metacarpal bone of the thumb, and into the os trapezium (Theile).

It moves the thumb out of the hollow of the hand towards the radius, hence the origin of the term abductor which has been given to it. After the thumb is drawn backwards, the muscle will assist in the extension of the radial side of the wrist.

The *extensor primi internodii pollicis*, C, is the smallest of the extensors, and arises from one bone and the interosseous membrane—being attached to the posterior surface of the radius for about an inch, but to rather more of the membrane. After passing through the annular ligament, the muscle is inserted into the base of the nearest phalanx; its tendon is united often with that of the extensor of the metacarpal bone.

Its primary action is to extend the nearest joint of the thumb; and contracting still more, the muscle will extend the wrist-joint.

Extensor secundi internodii pollicis, D, arises, like the preceding, from only one bone and the interosseous membrane, and chiefly from an impression on the ulna about four inches long, which lies inside that for the extensor ossis metacarpi. Contained in a separate space in the annular ligament, the tendon is continued over the back of the wrist, and the radial extensors of that joint, to its insertion into the base of the last phalanx of the thumb.

The muscle will extend the last joint of the thumb; and it can afterwards assist the other extensors in moving backwards the thumb, and extending the wrist.

Extensor indicis, E. The indicator muscle arises, inside the preceding, from the shaft of the ulnar for three or four inches below the middle (in length), though reaching sometimes as high as the anconeus. Passing through the annular ligament with the common extensor, it is directed to the fore finger, where it joins on the first phalanx the common tendinous expansion (p. 93).

The name expresses its action on the fore finger. If all the fingers are opened together, it assists the common extensor. When the fore finger is straightened, the other digits being closed, this muscle alone points the finger; for, during the act, the part of the common extensor to that finger is passive, being drawn out of the line towards the second finger by the fibrous band connecting the two outer pieces of the extensor tendon.

Supinator longus, M. In this Plate the peculiar shape of the upper part of the muscle, and the way in which it curves over the long extensor of the wrist to touch the short extensor, can be observed.

The *posterior annular ligament*, S, is described with Plate xi. In the dissection the sheath containing the common extensor of the digits was opened to trace the ending of the posterior interosseous nerve on the back of the wrist.

ARTERIES AT THE BACK OF THE FOREARM.

The posterior interosseous artery, and the ending of the anterior interosseous artery and some of its offsets, ramify amongst the muscles on the back of the forearm. Opposite the elbow joint the radial recurrent artery is directed backwards to the superficial muscles.

- *a*. Posterior interosseous artery.
- *b*. Recurrent interosseous.
- *c*. Communicating branch to the anterior interosseous.
- *d*. Continuation of the posterior interosseous artery.
- *e, e*. Perforating offsets of the anterior interosseous.
- *f, f*. Terminal parts of the anterior interosseous.
- *g*. Recurrent radial artery.
- *h*. Trunk of the radial artery.

The *posterior interosseous artery* springs from the common interosseous trunk in front of the 1 *r* b (p. 76, Plate ix.), and bends back above the interosseous membrane. Appearing, behind, between the supinator, A, and extensor ossis metacarpi, B, it is directed between the superficial and deep strata of the muscles as far as the lower third of the forearm: here it becomes superficial, and courses along the tendon of the extensor carpi ulnaris, G, to the wrist, where it ends in offsets, which communicate with the anterior interosseous, *f*, and with the posterior carpal, *b* (Plate xi.). Its named branches are recurrent and muscular.

Muscular branches supply the deep layer, and the digital and ulnar extensors of the superficial layer; those to the superficial layer have been cut in detaching the muscles.

The *recurrent branch*, *b*, ascends between the supinator, A, and anconeus, II; and supplying both muscles, anastomoses with the superior profunda artery. (Plate vii.)

The *anterior interosseous artery*, *f*, comes from the front, through an aperture in the lower part of the interosseous membrane, and ends on the back of the wrist, anastomosing with the posterior carpal and interosseous arteries; it gives a considerable offset to the outer side of the wrist.

Perforating branches of the anterior interosseous artery, *e, e,* three or four in number, pierce the interosseous membrane, and anastomose together as well as with the ending of the anterior interosseous, *f*.

Recurrent artery, g, of the radial, ascends beneath the supinator longus, M, and communicates with the upper profunda in the arm. (Plate vii.) It supplies the supinator, and the radial extensors of the wrist, also the brachialis anticus; and a considerable offset enters the supinator brevis, A, and communicates with the recurrent interosseous.

Radial artery, h. The anatomy of the trunk and branches of this artery on the back of the wrist and hand has been given in the description of Plate xi., to which reference may be made.

NERVE OF THE BACK OF THE FOREARM.

The musculo-spiral nerve supplies the extensor and supinator muscles of the back of the forearm.

1. Musculo-spiral trunk.
2. Radial nerve.
3. Posterior interosseous.
4. Branch to the two first extensors of the thumb.
5. Branch to the third extensor of the thumb and the indicator muscle.
6. Continuation of the posterior interosseous nerve.
7. Gangliform enlargement of the nerve on the wrist.

The trunk of the *musculo-spiral nerve*, 1, has been traced through the triceps to the outer part of the arm. (Plate vii.) Guided afterwards by the long supinator, M, and resting on the brachialis, N, it reaches the outer condyle of the humerus, and divides into two—radial and posterior interosseous. Branches from it enter the two muscles mentioned, also the long extensor of the wrist, and sometimes the short extensor.

The *radial nerve*, 2, has solely a cutaneous distribution, and ends in the integuments of the back of the hand, and the three outer digits.

The *posterior interosseous nerve*, 3, pierces the supinator brevis, and runs between the two strata of muscles to the middle of the forearm. Then sinking under the extensor of the second phalanx of the thumb, it is continued on the interosseous membrane to the back of the wrist, where it swells into a reddish gangliform body, 7, under the tendons of the common extensor, and gives offsets to the articulations.

All the muscles of the deep layer, and those of the superficial layer, except these three anconeus, long supinator, and long radial extensor of the wrist—receive branches from this nerve.

As the nerve supplies the extensors and supinators of the forearm, injury or disease of it may be attended by paralysis of those muscles; and as the flexor and pronator muscles in front, supplied by different nerves (median and ulnar), would then be unopposed in their action, they would determine the position of the limb. Consequently, after the function of the nerve is destroyed, the hand would be pronated, the wrist bent, and the fingers semiflexed by the action of the anterior group of muscles on the joints. This state of the limb is seen in the colic of painters.

With the subjoined concise notice of the general arrangement of the muscles, vessels, and nerves of the arm, and of the similarity between the two limbs, the anatomy of the upper limb will be brought to an end.

The upper has its counterpart in the lower limb; and with the palm of the hand up, the front of the upper limb is represented by the back of the lower; and the opposite.

The movements of the joints have a close resemblance in the two members; but the scapula and radius, possessing special movements, are provided with some muscles which are not required in the buttock and the leg.

As all the joints in the upper limb bend forwards, the flexors occupy the anterior, and the extensors the posterior surface; contrary to their position in the lower limb on the opposite aspects of each segment.

The vessels have a ramified distribution in the limbs—the branches diminishing in size, and increasing in number towards the digits, in the same way as the bones.

The offsets of the artery, unobstructed by valves, join freely together, and form larger and more frequent anastomoses the nearer they approach the digits; and in this way provision is made for the onward course of the blood even when the trunks may be closed.

The veins are provided with valves, which prevent a backward flow of the blood in them; and they are also united by collateral branches, so that the circulating fluid, stopped in one vessel, may be carried upwards more or less perfectly by another channel. Besides the deep veins, which are more numerous than the arteries they accompany, superficial veins ramify in the subcutaneous fat: both sets join at intervals.

In both limbs the nerves divide and decrease in size, like the arteries; but the branches are very constant, and regular in their distribution: they seldom join each other, unless they are subcutaneous.

All the nerves of the upper limb, with the exception of a few in the integuments of the shoulder and inner side of the arm, come from the brachial plexus. Each of the larger nerves supplies muscles and integuments. The smaller ones end altogether in the muscles about the shoulder. And two (large and small internal cutaneous) belong solely to the teguments.

Three nerves reach the fingers:—of these, one (musculo-spiral) ends on the dorsum; and the other two (median and ulnar) ramifying on the palmar surface of the digits, constitute specially the nerves of touch.

The three large nerves last mentioned supply most of the muscles below the shoulder:—the musculo-spiral being distributed to the extensors and the supinators, and to one flexor in part (brachialis anticus); and the ulnar and median giving branches to the flexors and the pronators.

ILLUSTRATIONS OF THE HEAD AND NECK.

DESCRIPTION OF PLATE XIII.

The base of the skull, with the cranial nerves, and the first and second stages of the dissection of the orbit, may be studied with the aid of this Figure.

After the removal of the brain, the fossæ and the dura mater in the base of the skull are visible without further preparation; but the dissection required for the display of the cranial nerves and the contents of the orbit will be subsequently described.

BASE OF THE SKULL AND THE DURA MATER.

The region called base of the skull is situated inside the cranium, and lies below the level of a line carried circularly round the head from the superciliary eminences in front to the occipital protuberance behind. It is divided into three fossæ on each side of the middle line; and a strong fibrous membrane, the dura mater, lines the whole.

A. Middle fossa of the base.
B. Posterior fossa.
C. Superior occipital fossa.
D. Part of the tentorium, cut through.
E. Part of the falx cerebri, also cut.
F. Falx cerebelli.
G. Straight sinus.
H. Cribriform plate of the ethmoid bone.
I. Crista galli of the ethmoid bone.
K. Roof of the orbit raised.

The *anterior fossa* of the base lies over the orbit, and must be destroyed nearly altogether by the dissection of that space. For the most part the surface of the fossa is convex, but along the middle line it is hollowed where it lodges the olfactory bulb: at the forepart of the hollow, H, small apertures exist in the cribriform plate of the ethmoid bone for

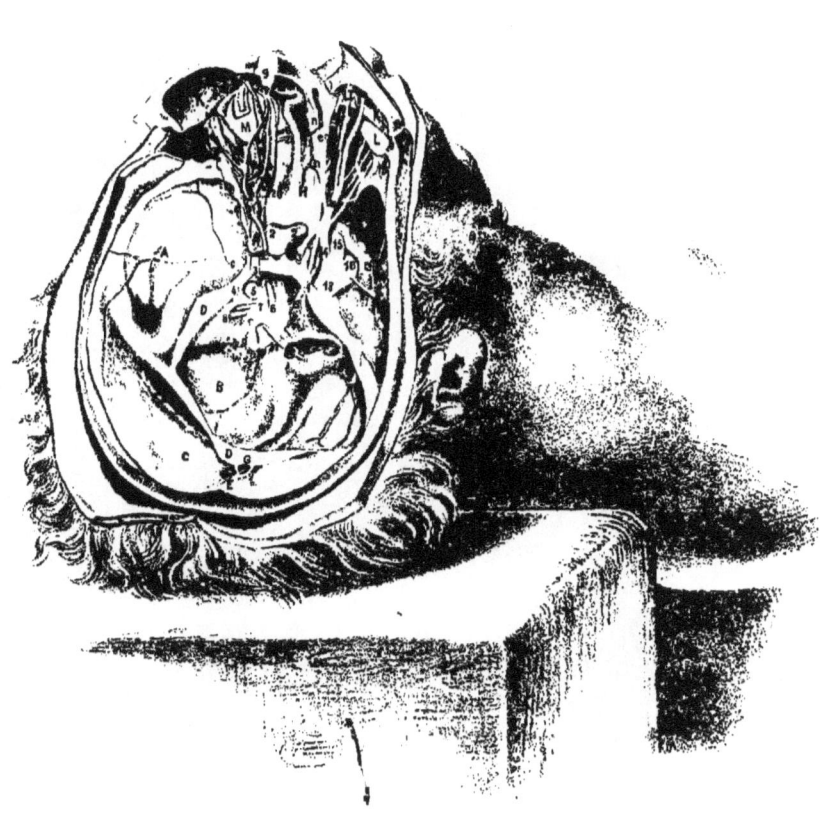

the transmission of the olfactory nerve filaments to the nose. On the anterior fossa rest the frontal lobes of the large brain.

The *middle fossa*, A, receives the middle part of the cerebrum or large brain. Much deeper than the anterior fossa, its bottom will reach down to a level with the articulation of the lower jaw. Along the middle line is the depression (sella Turcica) on the body of the sphenoid bone containing the pituitary body. Small vessels ramify in the fossa; and the internal carotid artery and some cranial nerves cross the inner end.

The *posterior fossa*, B, is more extensive than the others, being wide and shallow, and contains the hemispheres of the small brain or cerebellum. Its depth will be marked on the side of the head by the tip of the mastoid process. In its centre is the large foramen magnum transmitting the spinal cord.

If the skull has not been sawn so low as the occipital protuberance, there will be another depression at the base, the *superior occipital fossa*, C, in which the posterior end or the occipital lobe of the large brain rests.

Dura mater. This is a strong fibrous membrane, which serves as an endosteum to the bone, and supports parts of the brain. Its vessels and nerves are named meningeal. Three chief processes project inwards between parts of the brain: two of these are met with in the examination of the base of the skull, and the third occupies the middle line of the head above the occipital protuberance.

The *tentorium cerebelli*, D (Plate xiv.), is arched over the posterior fossa of the base, leaving only a small aperture in front for the communication of the spinal cord with the brain; and it is interposed between the large and small brains. Uniting with it along its middle, above and below, are folds—the falx cerebri and falx cerebelli, which keep it fixed tightly. In its centre is a triangular venous space, the straight sinus, G.

The *falx cerebelli*, F, reaches from the occipital protuberance to the foramen magnum, and is widest where it joins the tentorium. It contains the occipital sinus.

Falx cerebri, E. Only a small part of this is exhibited. It is narrowed in front and widened behind, and reaches along the middle line of the head from the crista galli, I, to the occipital protuberance where it joins the tentorium (Plate xiv.). At its attachment to the skull lies a venous space, the superior longitudinal sinus (Plate xiv. O).

Meningeal arteries. Small in size and few in number, they ramify in the dura mater of the fossæ, taking the names anterior, middle, and posterior, from their situation. Few of them are seen in an ordinary injection and they will be noticed more fully after the cranial nerves.

Meningeal nerves. These are smaller than the arteries, and cannot be perceived without steeping the dura mater in acid: they are derived from the sympathetic, and from some of the cranial nerves, especially the fifth.

CRANIAL NERVES IN THE BASE OF THE SKULL.

All the nerves attached to the encephalon are called cranial; and one nerve, 11, not attached to the encephalon, is reckoned as a cranial nerve, because it enters the skull and leaves by an aperture in the base of the cranium. The nerves course forwards from their origin to the apertures of exit; and a part of each nerve is left in the skull after the removal of the brain.

The nerves crossing the middle fossa of the base of the skull are invested by sheaths of the dura mater, but the others are free from the same till they enter their foramina of exit. On the left side, the place of entrance of those nerves into the sheaths may be observed; but to examine fully their trunks, and to define also the ganglion and branches of the fifth nerve, as in the Figure, the dura mater should be removed on the right side from the middle fossa of the base.

There are twelve pairs of cranial nerves:*—these are marked by corresponding numerals, except in the case of the first nerve which has been removed with the brain.

2. Optic nerve and commissure.
3. Motor nerve of the eyeball.
4. Trochlear nerve.
5. Trifacial nerve.
6. Abducent nerve of the eyeball.
7. Facial nerve.
8. Auditory nerve.
9. Glosso-pharyngeal nerve.
10. Pneumogastric nerve.
11. Spinal accessory nerve.
12. Hypoglossal nerve.
13. Gasserian ganglion.
14. Ophthalmic nerve.
15. Superior maxillary nerve.
16. Inferior maxillary nerve.
17. Large petrosal nerve.

---

* English anatomists reckon in general nine pairs of cranial nerves, and the anatomists on the Continent enumerate twelve pairs; so that some confusion in the nomenclature arises from this difference in the mode of numbering. The enumeration of the nerves as twelve appears most natural, as only two nerve

The *olfactory*, or *first cranial nerve*, is marked by a bulb which rests on the cribriform plate of the ethmoid bone, and sends filaments to the nose through the subjacent apertures: it will be found attached to the brain.

The *optic*, or *second nerve*, 2, ends in the eyeball. Posteriorly the nerves of opposite sides unite in a commissure (chiasma) on the olivary eminence of the sphenoid bone, with a partial decussation of their fibres. In front the nerves diverge; and each issues from the skull through the optic foramen, with the ophthalmic artery. In the orbit of the left side the further course of the nerve to the eyeball is evident.

The *motor oculi*, or *third nerve*, 3, crosses the middle fossa, and enters its sheath of dura mater behind the anterior clinoid process, as seen on the left side. Contained in the dura mater, it is conveyed to the sphenoidal fissure, and supplies all the muscles moving the eyeball, except two.

The *trochlear*, or *fourth nerve*, 4, is received into sheath of dura mater behind the posterior clinoid process, and courses forwards through the wall of the cavernous sinus to end in one muscle in the orbit—superior oblique.

The *trifacial*, or *fifth nerve*, 5, consists of two roots, large and small, though only the large root is visible, for this lies over and conceals the small root.

The *large root* enters a sheath of dura mater above the petrous portion of the temporal bone, and swells into a large ganglion in the middle fossa of the skull.

This ganglion, 13, named Gasserian, and about as large as the thumb-nail, is widened in front, and is crossed by a ridge to which the dura mater adheres closely. From the fore part of the ganglion three large trunks are sent forwards to end on the face, hence the origin of the name of the nerve:—the highest of these is the ophthalmic trunk, 14, which passes through the sphenoidal fissure to the orbit; the middle one, or the superior maxillary, 15, leaves the skull by the foramen rotundum; and the third, the inferior maxillary nerve, 16, issues from the skull through

---

trunks, with like function and distribution, will then be included in one cranial pair; whilst, in using the smaller number, as many as four and six trunks, differing in name, function, and distribution, will be combined together as one pair of the cranial nerves.

the foramen ovale. These trunks of the ganglion confer sensibility on the parts to which they are distributed.

The *small* root of the fifth lies under the large one, and will come into view on raising the ganglion; it is unconnected with the ganglion, and belongs exclusively to the inferior maxillary trunk. Blending with offsets of the inferior maxillary trunk outside the skull, it is conveyed to muscles, and chiefly to those of mastication, to which it gives the power of contracting.

The *abducent,* or *sixth nerve,* 6, pierces the dura mater behind the body of the sphenoid bone, and entering the cavernous sinus, passes through the sphenoidal fissure to one muscle (external rectus) of the orbit.

All the nerves crossing the middle fossa of the base of the skull, viz., the third, fourth, fifth, and sixth, communicate with the sympathetic on the carotid artery.

The *facial,* or *seventh nerve,* 7 (portio dura of the seventh pair, Willis), enters the meatus auditorius internus. In the bottom of that hollow it is received into the aqueduct of Fallopius, and is conveyed to the stylo-mastoid foramen, where it escapes, to be distributed to the muscles of the face, the head, and the ear (in part); it is the motor nerve of those muscles.

The *auditory,* or *eighth nerve,* 8 (portio mollis of the seventh pair, Willis), soft, and divided into fibrils, accompanies the facial into the meatus auditorius, and passes through the small apertures in the bottom of that meatus, to end in the inner parts of the ear.

The *glosso-pharyngeal,* or *ninth nerve,* 9 (part of the eighth pair, Willis), leaves the skull by the foramen jugulare, being contained in a distinct sheath of dura mater, and lying in a depression in the lower border of the temporal bone. It is distributed, as the name expresses, to the tongue and pharynx.

The *pneumogastric,* or *tenth nerve,* 10 (part of the eighth pair, Willis), is transmitted through the foramen jugulare in a sheath of dura mater common to it and the following nerve. It is a flat trunk, consisting of fibrils. Its terminating branches ramify in the air passages, the heart, and the alimentary canal.

The *spinal accesory,* or *eleventh nerve,* 11 (part of the eighth pair, Willis), is the only cranial nerve that is not united with the encephalon. Arising from the spinal cord, it enters the skull through the foramen magnum; it then bends outwards to the foramen jugulare, and leaves the cranium through that hole in close contiguity to the pneumogastric—the two being contained in the same fibrous sheath. This nerve supplies in part two muscles of the neck—the sterno-mastoid and trapezius.

The *hypoglossal*, or *twelfth nerve*, 12 (ninth pair, Willis), consists of two bundles of filaments, which pierce separately the dura mater. These join in the anterior condyloid foramen, by which they issue from the cranium as one trunk. It is a motor nerve of some of the hyoid, and the tongue muscles.

Large petrosal nerve, 17. This is a continuation of the Vidian nerve, derived from Meckel's ganglion. Coming into the skull through the pterygoid foramen and over the foramen lacerum in the base, it is conveyed in a bony groove under the Gasserian ganglion to the hiatus Fallopii, which it enters to join the facial nerve in the temporal bone.

VESSELS IN THE BASE OF THE SKULL.

Two large arteries on each side, carotid and vertebral, pass through the base of the skull in their course to the brain, and furnish some offsets to the dura mater. Other meningeal vessels, supplied from arteries outside the cavity of the skull, ramify in the dura mater.

a. Internal carotid artery.	m. Posterior meningeal artery.
b. Vertebral artery.	n, n. Anterior meningeal arteries.
c. Large meningeal artery.	

The *internal carotid artery*, a, issues from the carotid foramen in the apex of the temporal bone, and winding through the cavernous sinus (Plate xiv.), touches the brain at the inner end of the fissure of Sylvius, and splits into branches (cerebral) for the supply of the large brain or cerebrum. On the side of the sphenoid bone it makes two bends, lying internal to the cranial nerves; and at the base of the brain it is placed between the second and third nerves.

An ophthalmic branch, and small offsets to the dura mater, spring from this part of the carotid.

The *vertebral artery*, b, is a branch of the subclavian trunk, and enters the skull through the foramen magnum: the arteries of the opposite sides soon coalesce, and they supply the small, and part of the large brain. An offset is furnished by it to the dura mater.

Meningeal arteries. Small arteries ramify in each fossa of the base of the skull; they are named anterior, middle, and posterior, like the fossæ.

The *anterior set*, two in number, n, n, and the smallest, are branches of the ophthalmic artery in the orbit: they come from the anterior and posterior ethmoidal arteries, and entering the skull at the edge of the cribriform plate, end in the middle part of the fossa. One sends a twig to the front of the falx cerebri, E.

The *middle set*, three in number are derived from branches of the external carotid artery, and appear through the lacerated, oval, and spinous foramina. The largest of these, and the only one generally injected is the following:—

The middle meningeal artery, c, nourishes chiefly the bony case containing the brain. Arising from the internal maxillary artery, it comes inwards through the foramen spinosum, and ascends to the top of the head, grooving the bones—more particularly the lower and fore parts of the parietal. At the vertex of the skull it terminates in the bone, but some branches communicate with the arteries on the outer surface of the cranium.

Branches are given by it to the dura mater. A petrosal branch enters the hiatus Fallopii with the small nerve, 17, to supply the temporal bone; and one or two offsets penetrate into the orbit, and join the ophthalmic artery.

The *posterior set* includes two arteries: one is furnished by the occipital through the foramen jugulare, and the other, by the vertebral artery inside the skull. Of the two, the offset, m, from the occipital is the largest, and it extends even to the tentorium cerebelli.

Veins. No vein accompanies either the internal carotid or the vertebral vessels which end in the brain; but companion veins run with the arteries distributed to the dura mater and the brain case. The veins with the large middle meningeal artery may be plainly seen in a dissection.

CONTENTS OF THE ORBIT.

In the orbit is lodged the eyeball with its muscles, vessels, and nerves. And the gland for the secretion of the tears is contained in the fore part of the same cavity.

The dissection of this cavity requires some care in its execution, in consequence of the smallness of the vessels and nerves, and of the quantity of fat with which they are surrounded.

On the right side the first stage of the dissection has been prepared by sawing through and throwing forwards the bony roof; and then slitting along the middle, and removing the periosteum of the cavity. On the left side, the cavity having been opened as before, the superficial layer has been taken away, to bring into view deeper vessels and nerves.

SUPERFICIAL MUSCLES AND THE LACHRYMAL GLAND.

The muscles contained in the orbit act on the eyeball, with the exception of one which raises the upper eyelid. Six muscles are attached to the eyeball; of these, four are straight, and direct the pupil to opposite points of the circumference of the orbit; whilst two, which are named oblique, roll the ball.

L. Lachrymal gland.
M. Eyeball of the left side.
N. Upper oblique muscle.
P. Levator palpebræ superioris.

R. Upper rectus muscle.
S. External rectus muscle.
T. Pulley of the upper oblique muscle.

The *lachrymal gland*, L, lies above the muscles in the outer part of the orbit, and touches in front the upper eyelid. Shaped somewhat like an almond, with its longest measurement directed transversely, it is suspended by fibrous tissue to the roof of the orbit. It secretes the tears; and its ducts, six or eight in number, open along an arched line on the inner surface of the upper lid, near the outer end.

The *upper oblique muscle*, N (trochlearis), is the longest muscle in the orbit, and passes through a ring, or pulley. It arises from the frontal bone, close to the optic foramen in the bottom of the orbit; and ends anteriorly in a tendon, which is directed backwards through the pulley, but beneath the upper rectus, and is inserted into the eyeball behind the centre (Fig. xiv.).

The trochlea or pulley, T, is a ring of fibro-cartilage, which is attached to the pit near the inner angle of the frontal bone. A synovial membrane lines the ring, and fibrous tissue is prolonged from the margins along the tendon.

The muscle draws inwards somewhat the back of the eyeball, rotating it at the same time time; and it gives to the pupil an inclination downwards and outwards towards the top of the shoulder. By this action it is thought to control the movement downwards and inwards of the eye by the inferior rectus muscle.

The *levator palpebræ superioris*, P, arises in the bottom of the orbit, close to the preceding; becoming tendinous in front of the eyeball, it enters the upper eyelid, and is inserted into the fore part of the tarsal cartilage.

The muscle elevates the upper eyelid, moving upwards the fibro-cartilage over the eyeball, and gives rise to a deep groove in the skin. If the eyeball is directed down when the muscle is acting, the elevation of the lid is checked by the mucous membrane which is then less loose.

Recti muscles. The upper rectus, R (attollens oculi), and the outer rectus, S (abductor oculi), have a common origin with the other two recti, around the optic nerve, at the apex of the orbital cavity; and they are inserted into the eyeball about a quarter of an inch behind the cornea.

The outer rectus is provided with an additional origin from a point of bone on the lower edge of the sphenoidal fissure, near the inner end of that slit: between this head and the common one the ophthalmic vein and several nerves pass.

The pupil is directed upwards and inwards by the upper rectus muscle, and outwards by the other rectus—the insertion of the muscles into the ball in front of its greatest transverse diameter impressing on the eye the movements stated. Squinting upwards or outwards may result from permanent contraction of the muscle moving the eye in the direction indicated, or from the rectus in action being unbalanced through paralysis of its antagonist muscle or muscles.

VESSELS OF THE ORBIT.

The ophthalmic artery and vein are represented in the left orbit. These vessels have some peculiarities:—they are not transmitted through the same aperture in the bone; and the vein, which is a single trunk, ends in the cavernous sinus in the interior of the skull.

d. Ophthalmic artery.
e. Anterior ethmoidal or nasal artery.
f. Posterior ethmoidal artery.
g. Supra-orbital artery.
h. Ophthalmic vein.

The *ophthalmic artery, d,* is a branch of the internal carotid, and enters the orbit through the optic foramen, lying below and outside the optic nerve. In the left orbit the artery is shown coursing over the optic nerve, and along the inner side to the front of the cavity, where it ends in branches for the root of the nose (nasal) and the forehead (frontal). Most of its offsets are distributed in the orbit.

Offsets for the eyeball. Several branches, *posterior ciliary,* pierce the back of the eyeball around the optic nerve. Other smaller arteries, which are usually not injected, enter the front of the ball, close to the cornea: these are the *anterior ciliary,* and they are best seen in inflammation of the iris. One artery enters the optic nerve behind the ball; it ramifies in the retina, and is called the *central artery of the retina.*

The *lachrymal branch* accompanies the nerve, 19, to the gland of the same name.

Muscular branches arise at intervals: some of these are seen in the Figure.

Eyelid offsets. Each eyelid receives a *palpebral* branch: these are directed transversely, in contact with the tarsal cartilages, and anastomose externally with the lachrymal—forming an arch in each lid.

Branches leaving the orbit. Besides the two terminal branches (frontal and nassal), three others leave the cavity. One *anterior ethmoidal, e,* accompanies the nassal nerve to the nose, and supplies meningeal offsets. Another, *posterior ethmoidal, f,* smaller than the preceding, passes through the foramen of the same name to the dura mater in the anterior fossa of the skull. And the third, *supra-orbital, g,* runs with the nerve of the same name through the supra-orbital notch to the forehead.

The *ophthalmic vein, h,* taking the same general course as the artery, joins in front the facial vein; and as its branches correspond mostly with those of the artery few are delineated. At the back of the orbit it leaves the artery, and passing between the heads of the outer rectus, ends in the cavernous sinus in the skull (Plate xiv. Q).

Eyeball veins:—These differ from the arteries of the ball in their number and course. Four in number, they issue on opposite sides of the eye, and about midway between the cornea and the entrance of the optic nerve.

8

NERVES OF THE ORBIT.

Five cranial nerves enter the orbital cavity, viz. 2d, 3d, 4th, 5th, (in part) and 6th; and all, except the second or optic, come through the sphenoidal fissure. Some end in the contents of the orbit, like the arteries, and others are transmitted through the cavity to the nose and the forehead: they have the following general distribution. The second or the optic belongs to the eyeball. The third, fourth, and sixth, are furnished to muscles. And the ophthalmic trunk of the fifth nerve supplies the eyeball and the lachrymal gland, and ends outside the orbit.

The nerves which are superficial to the muscles are displayed on the right side, viz. the fourth, and the supra-orbital and lachrymal branches of the fifth: on the left side the other nerves referred to in the description may be observed.

2. Optic nerve.	18. Supra-orbital nerve.
3. Third nerve.	19. Lachrymal nerve.
4. Fourth nerve.	21. Upper branch of the third nerve.
14. Ophthalmic nerve of the fifth.	23. Continuation of the nasal nerve.

The *third cranial nerve*, 3 (motor oculi), supplies all the muscles of eyeball except two, and enters the orbit in two pieces between the heads of the external rectus. The upper and smaller part, 21 (left side), is furnished to the levator palpebræ, P, and the upper rectus, R; the lower portion of the nerve may be seen in Plate xiv. 22.

The *fourth cranial nerve*, 4, passes through the sphenoidal fissure above the muscles, and ends in the upper oblique, X, piercing the fibres of the muscle on the surface turned away from the eyeball.

The *ophthalmic nerve*, 14, begins in the Gasserian ganglion, 13, and is continued through the wall of the cavernous sinus and the sphenoidal fissure to the orbit. It ends by dividing into the supraorbital, 18, and the lachrymal branch, 19; and from its inner side, before the terminal bifurcation, springs the nasal nerve, 20 (Fig. xiv.).

The *lachrymal nerve*, 19, the smallest of the offsets of the ophthalmic trunk, is directed to the outer part of the orbit, and supplies the lachrymal gland and the upper eyelid.

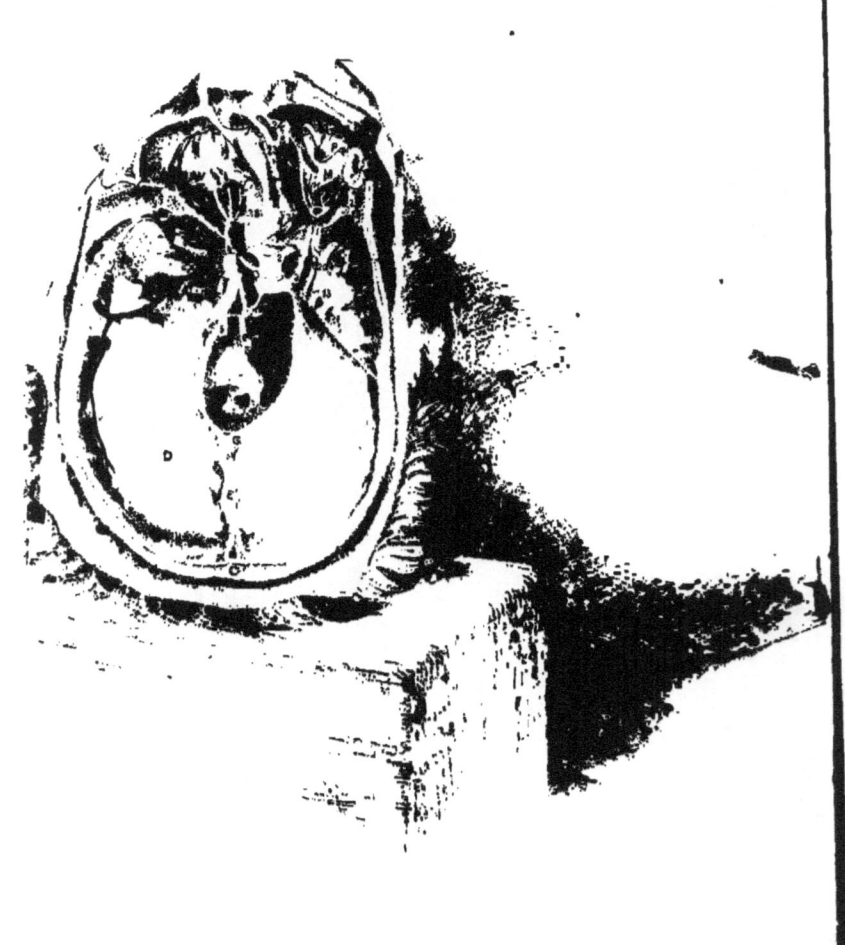

The *supra-orbital nerve*, 18, lies above the muscles, like the lachrymal, and is continued through the cavity to the supra-orbital notch, where it issues on the forehead, and supplies the muscles and the integuments. From its inner side is given a long slender branch, *supra-trochlear*, to the upper eyelid and the forehead; and as it turns round the margin of the orbit, small *palpebral* filaments are furnished to the upper eyelid.

The *nasal nerve* (20, left side) ends in the nose, and passes through the orbit and the cavity of the skull before it reaches its destination. Entering the orbit between the heads of the external rectus (Plate xiv.) it is continued forwards with the ophthalmic artery to the anterior of the two foramina in the inner wall (23, left side); here it is transmitted to the cavity of the skull. In the orbit its offsets are the following:—

Firstly, there is a slender communicating branch to the lenticular ganglion (Plate xiv. 26). As it crosses the optic nerve two or three filaments, *long ciliary*, are furnished to the eyeball. And as it leaves the orbit it gives a branch—*infra-trochlear*, 24, to the upper eyelid and the side of the nose.

The nasal nerve is distributed finally to the mucous membrane of the front of the nasal cavity, and to the integuments of the end of the nose. Irritation of it in the nasal cavity, as in taking snuff, induces sneezing for the purpose of removing the irritating body.

Sixth cranial nerve, 6. The ending of this nerve in the external rectus is delineated in Fig. xiv.

DESCRIPTION OF PLATE XIV.

In the Drawing a view is obtained of the dura mater at the base of the skull, with the cavernous sinus; and the dissection of the orbit is carried through its two deeper stages.

Parts delineated in this and the preceding Plate are marked by the same letters and figures.

THE DURA MATER WITH THE SINUSES.

The *tentorium cerebelli*, D, has been left entire for the purpose of showing the height and extent of this partition. Its position will be marked on the surface by a line on a level with the part of the ear joining the side of the head.

Venous spaces occupy the middle part, and the attached edge of the membrane; and one of the largest spaces, called the cavernous, is close to each anterior extremity.

The *cavernous sinus*, Q, may be opened, as on the left side, by cutting through the dura mater from the anterior clinoid process to the petrous portion of the temporal bone, the cut being made internal to the third and fourth nerves.

This hollow is placed on the side of the body of the sphenoid, and reaches from the sphenoidal fissure to the temporal bone. Rather more than an inch long, it measures across about half an inch, after the handle of the knife has been put into it; and it is dilated behind where it joins other sinuses. Its inner boundary is formed by the sphenoid bone covered by thin dura mater; and the outer boundary, consisting of thickened dura mater, contains the third, 3, fourth, 4, and the ophthalmic trunk of the fifth nerve, 14, Plate xiii.

Through the inner part of the space pass the internal carotid artery and the sixth cranial nerve; and these are separated from the blood by the thin venous lining membrane. Small fibrous bands and arteries traverse the space, giving rise to the term "cavernous."

Blood is received from a few small cerebral veins which pierce the outer wall, though chiefly from the ophthalmic vein (Fig. xiii. *h*) which enters in front; and it circulates backwards to be conveyed to the lateral sinus by the upper and lower petrosal sinuses. The blood in the space communicates with that outside the head by means of small emissary veins, which penetrate through the foramen lacerum.

Three short sinuses join the cavernous spaces of opposite sides across the middle line;—one lying before the pituitary body, one behind it; and the other across the basilar process of the sphenoid bone. No valves exist in these cross channels, so that the blood can move freely forwards and backwards through them; and should the diminished size or the absence of one lateral sinus interfere with, or stop the course of the blood on that side of the skull, the circulating fluid can be conveyed across the middle line, to be transmitted from the head by the lateral sinus of the opposite side.

The *internal carotid* artery, *a,* winds through the space from behind forwards, and issues through the dura mater internal to the anterior clinoid process: it furnishes here small *receptacular* branches to the dura mater.

Ascending around the artery is the cranial part of the *sympathetic nerve,* which communicates with the nerves entering the orbit through the sphenoidal fissure.

In the sinus lies the *sixth cranial* nerve, 6, which courses from behind forwards, close outside the carotid artery, and communicates largely with the sympathetic.

Another large central sinus, named *torcular Herophili,* is placed opposite the occipital protuberance, and receives blood from the brain. Opening into it in front is the straight sinus G (Plate xiii.); above is the superior longitudinal, O; and below is the occipital sinus contained in the falx cerebelli. On each side issues the large lateral sinus, which extends to the foramen jugulare, joining there the internal jugular vein, and conveys from the skull the blood both of this and of the cavernous sinus.

DISSECTION OF THE ORBIT.

The third stage of the dissection, which is represented on the right side, will be obtained by clearing away the vessels shown in the left orbit in Plate xiii. And the preparation of the last stage, as exhibited on the left side, may be made by removing the lenticular ganglion and the nasal nerve, and by dividing the optic nerve and raising the ends.

MUSCLES OF THE ORBIT.

The muscles lying below and to the inner side of the eyeball act as antagonists to the group of muscles (before described, p. 111) on the outer side and above the ball. Like the other group they consist of two straight and one oblique; and they are named inferior rectus, internal rectus, and inferior oblique.

N. Upper oblique muscle.
O. Superior longitudinal sinus.
P. Levator palpebræ superioris.
Q. Cavernous sinus.
R. Upper rectus muscle.

S. External rectus muscle.
V. Inferior rectus muscle.
W. Inferior oblique muscle.
X. Internal rectus muscle.

Recti muscles. The lower rectus, V (depressor oculi), and the inner rectus, X (adductor oculi), arise, behind, around the optic nerve with the other muscles; and the two separating from each other in front, are inserted into the eyeball near the cornea, each being attached opposite its antagonist muscle.

One of these muscles contracting, the pupil will be directed towards it, the under rectus depressing and adducting, and the inner one adducting the eye; but the two recti acting together the pupil will be turned to a point intermediate between both.

The *external rectus*, S, is more evident here than in Figure xiii.; and on the right side the nerves passing between its heads of origin, viz., the third, 3, the nasal nerve of the fifth, 20, and the sixth, 6, have been traced out, to show their relative position.

The *inferior oblique muscle*, W, is displayed only at its insertion into the eyeball. Arising from the fore part of the floor of the orbit, close to the lachrymal sac, it is inclined backwards below the inferior rectus and the eyeball, and is inserted into the back of the eye near the upper oblique muscle.

By the action of this muscle the back of the ball may be depressed and the cornea raised; and the eye being rotated at the same time the cornea will be directed upwards and outwards towards the temple. This movement towards the outer side of the orbit is thought to counteract the motion of the ball up and in by the upper rectus muscle.

DEEP NERVES OF THE ORBIT.

The second nerve, part of the third nerve, the lenticular ganglion, and the sixth nerve, are met with in the two deeper stages of the dissection of the orbit.

DEEP NERVES OF THE ORBIT.

On the right side the lenticular ganglion is depicted, with the optic nerve; and the other nerves are visible on the left side.

2. Optic nerve.	21. Upper branch of the third nerve.
3. Third cranial nerve.	22. Lower branch of the third nerve.
4. Fourth nerve.	23. Nasal nerve leaving the orbit.
5. Fifth cranial nerve.	24. Infra-trochlear nerve.
6. Sixth cranial nerve.	25 Lenticular ganglion.
13. Gasserian ganglion.	26. Long root of the lenticular ganglion to the nasal nerve.
20. Nasal nerve at its origin.	

The *optic* or *second cranial nerve*, 2, lies in the middle of the hollow included by the recti muscles, and enters the back of the eyeball rather internal to the centre: it spreads out in the nervous stratum of the retina. Along it the ciliary arteries and nerves are conveyed to the eyeball.

The *ophthalmic* or *lenticular ganglion*, 25, is a small, rather red body, about as large as a pin's head of moderate size, which is situate at the back of the orbit, close to the ophthalmic artery and the optic nerve. Nerves issue from it at four points (angles): two pass backwards, joining other nerves, and these are called roots; and several nerves are sent forwards to the eyeball along the optic nerve.

Posterior branches.—A long, slender branch—the long root, 26, joins the nasal nerve, 20. Another thick and short branch—the short root—unites with the third nerve, 22 (right side). Sometimes a third offset, between those two, connects the ganglion with the sympathetic.

The anterior branches or the short *ciliary* nerves to the eyeball, are about twelve in number, and form two bundles, upper and lower: they are furnished to the ball, and especially to the muscular structure in it.

The *third cranial nerve*, 3, splits into two as it is about to enter the orbit between the heads of the outer rectus. Its upper piece, 21, ends in the upper rectus, and in the elevator of the upper eyelid.

The lower and larger part of the nerve, 22 (left side), divides into three: one enters the inferior rectus, V; the second belongs to the internal rectus; and the third offset, 22 (right side), is continued below the eyeball to the inferior oblique muscle. The last branch is joined by the short root of the lenticular ganglion, and supplies through that communication motor nerves to the muscular fibres of the eyeball.

Paralysis of the muscles supplied by the third nerve is attended by dropping of the eyelid, and inability to raise it; and the eye loses its movements in certain directions. Supposing its existence on one side, the cornea could not be moved vertically, that is to say, it could not be turned upwards or downwards by the elevator and depressor muscles; it could not be drawn inwards horizontally by the adductor; nor could it be inclined upwards and outwards by the inferior oblique —all the muscles needful for those movements being supplied by the nerve, and being therefore unable to contract. Only two movements would remain, viz., abduction and rotation downwards and outwards: —the former depending on the external rectus which is supplied by the sixth nerve; and the latter, on the superior oblique, which receives the fourth nerve.

Double vision will accompany the paralysis when an attempt is made to look with both eyes to the temple of the opposite or healthy side; and this occurrence is to be accounted for by the loss of the muscular control over the ball of the affected side. In looking with both eyes to the temple (left) in the undiseased state of the muscles, the left eye will be inclined outwards by the external rectus, and the right eye will be turned inwards, towards its fellow, by the internal rectus. But in paralysis, say of the right side, the affected eye cannot be inclined towards its fellow in consequence of the internal rectus having lost its power of contracting, whilst the healthy or left eye will be turned outwards by the external rectus muscle; and as the axes of the eyes are not kept parallel, images are formed on non-corresponding points of the two retinæ, and double vision results.

The *sixth cranial nerve*, 6, enters the orbit between the heads of the external rectus, lying below the third and nasal nerves, and above the ophthalmic vein: it is distributed to the external rectus muscle.

In paralysis of the external rectus from disease of this nerve the eyeball cannot be directed outwards; and squinting inwards may ensue from the absence of a contracting muscle to balance the internal rectus.

Orbital branch of the *upper maxillary* nerve. After the contents of the orbit have been removed, this small nerve may be found in the lower and outer angle, passing through the orbit on its way to the face and the temple.

PLAT

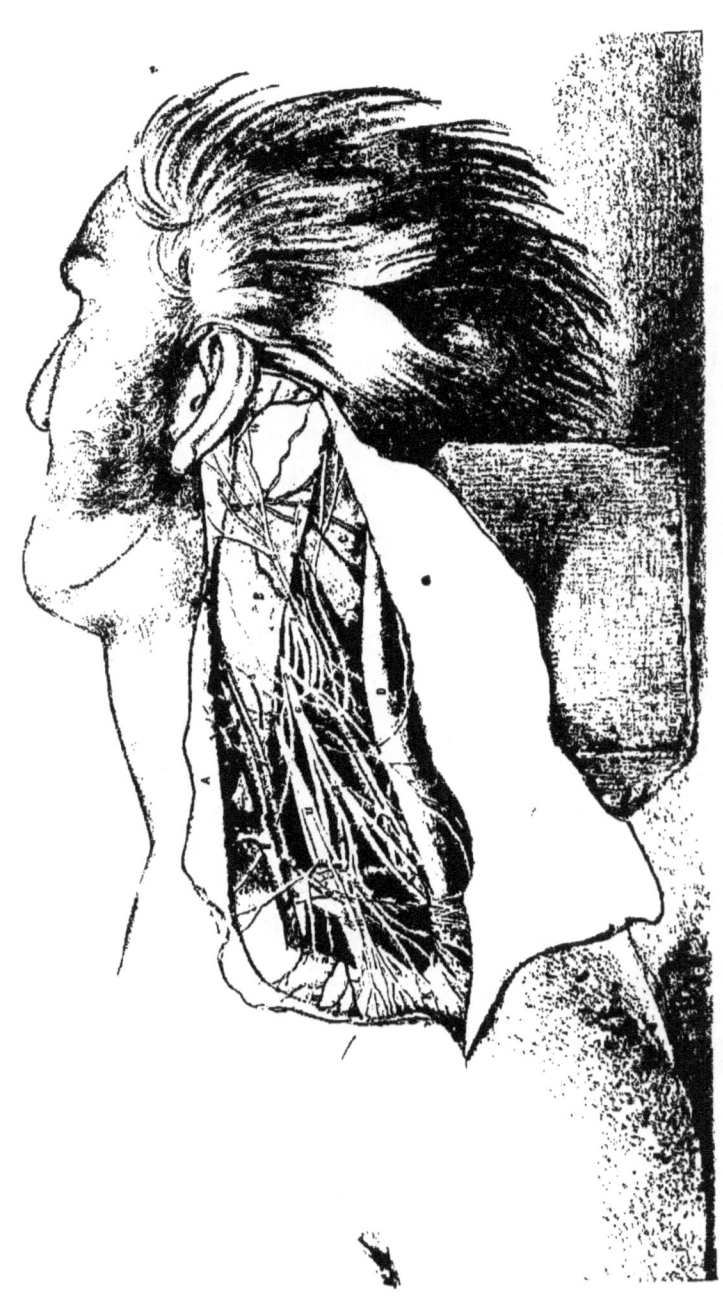

DESCRIPTION OF PLATE XV.

This Figure illustrates the anatomy of the side of the neck behind the sterno-mastoid muscle.

The position of the body indicated in the Drawing will be required also during the dissection, viz. the arm having been drawn down to depress the shoulder, and to make tense the neck muscles.

The more prominent lateral muscles will appear readily on reflecting the skin by the incisions marked in the Plate, and on removing the thin platysma muscle, and the deep cervical fascia; but much time and care will be needed to make clean, and to leave uninjured the deeper nerves and vessels.

MUSCLES OF THE SIDE OF THE NECK.

All the muscles here exhibited in part, are attached below either to the arch formed by the clavicle and the scapula, or to the first rib; and above they are fixed to the head and the spinal column, with the exception of the omo-hyoid which is attached to the hyoid bone. A hollow, the posterior triangular space, intervenes between the two largest superficial muscles.

A. Platysma myoides.
B. Sterno-cleido-mastoid muscle.
C. Splenius capitis.
D. Trapezius.
E. Levator anguli scapulæ.
F. Scalenus medius.

G. Scalenus anticus.
H. Omo-hyoideus.
K. Deltoid muscle.
L. Clavicle.
N. Pectoralis major.

Platysma myoides, A. This is a membraniform fleshy layer, which is contained in the fatty stratum between the skin and the deep fascia. Arising from the scapular arch, and the top of the thorax and shoulder, it crosses the side of the neck, and is inserted into the base of the lower jaw, blending with muscles of the face.

It covers the external jugular vein, k, and the lower two thirds of the posterior triangular space. Its fibres are inclined downwards and backwards from the jaw to the shoulder; and in opening the external jugular vein in venesection the incision is to be so directed as to cut them across.

The *sterno-cleido-mastoid muscle*, B, crosses the neck obliquely from the thorax to the ear. Below, it arises from the first piece of the sternum, and the inner third of the clavicle (Plate xvii.); and it is inserted above into the mastoid portion of the temporal, and the upper curved line of the occipital bone.

From its diagonal position in the neck it separates a triangular hollow in front from another behind: it covers the great carotid bloodvessels and the neck muscles, and is crossed by superficial nerves and veins. It is pierced by one large nerve—the spinal accessory or the eleventh cranial nerve, 13.

Both muscles acting, the head will be brought forwards, as in nodding, or the sternum will be raised; according as they may take their fixed point above or below. If only one muscle is used the head is turned to the opposite side; but in combination with other muscles attached to the mastoid process it can incline the head towards the shoulder on the same side.

In wry-neck (torticollis) arising from muscular contraction, the sterno-mastoid forms a hard, tense cord on the side of the neck opposite to that to which the head is turned. Subcutaneous cutting through of the muscle has been practised to remove the deformity.

The *trapezius*, D, attached behind to the spinal column and the head, is inserted in front into the outer third of the clavicle, and into the acromion process and the spine of the scapula.

The anterior free edge of the muscle limits behind the posterior triangular space; it is thin in the upper half, and it is projected forwards, as a point, opposite the fourth cervical nerve and the narrowed part of neck.*

The fore part of the trapezius will help the levator anguli scapulæ, E, in raising the shoulder.

Splenius capitis, C. This small part of the splenius muscle appears

---

* When this edge is represented in Anatomical Plates as straight between the upper and lower attachments, the displaced condition is delineated.

in the posterior triangular space, where it arches forwards from the spinal column to the mastoid process.

Taking its fixed point behind, it can turn the face to its own side; or acting with the sterno-mastoid, it will incline the head to the shoulder. When the muscles of opposite sides act together, the head will be carried backwards.

Levator anguli scapulæ, E, occupies the hinder part of the triangular space. It arises from the transverse processes of the three or four upper cervical vertebræ, and is inserted into the base of the scapula (Plate v. C.); its processes of origin may remain separate for some distance as in the Plate, and appear like distinct muscles.

Its ordinary action is manifested in shrugging the shoulders; in this movement it is assisted by the upper part of the trapezius.

The *omo-hyoideus* is a double-bellied muscle, which reaches from the scapula to the hyoid bone, and is tendinous beneath the sterno-mastoideus (Plate xviii.): for the anatomy of the anterior belly, see Plate xvii.

The fibres of the posterior belly, H, are attached beneath the trapezius to the upper border of the scapula, close to the notch in that bone; and they end in front in the intermediate tendon. This belly crosses the posterior triangular space, cutting off a small part below, which contains the subclavian artery; and it is kept in place by a sheath of the cervical fascia.

This belly of the muscle makes tense the deep fascia of the neck. The possibility of its compressing the internal jugular vein has been suggested by Theile.*

The *scaleni muscles*, three on each side, pass from the first two ribs along the side of the spinal column, and are crossed by the great nerves and vessels of the upper limb.

The *anterior muscle*, G, arises from the first rib around a slight prominence on the upper surface; and it is inserted into the fore part of the transverse processes of four cervical vertebræ, viz., 6, 5, 4, 3.

In front of the muscle lie the omo-hyoideus, H, and sterno-mastoideus, B; but the deep connections can be more fully observed in Plate xviii. With a lateral view of the side of the neck, as in the Figure, part of the muscle may be seen in the posterior triangular space; but in a front view, the muscle is usually concealed by the sterno-mastoideus.

---

* "Lehre von den Muskeln," Leipzig, 1841.

The *middle muscle*, F, larger than the preceding, arises from a groove across the hinder part of the upper surface of the first rib; and it is inserted into the posterior part of the transverse processes of all the cervical vertebræ.

Along its outer edge lies the levator anguli scapulæ; and it is placed beneath the cervical nerve trunks, and the subclavian artery.

The *posterior muscle* is small, and is concealed by the preceding. Arising from the upper border of the second rib at the back, it is inserted into the transverse processes (posterior or neural) of two or three lower cervical vertebræ.

When the neck is fixed the scaleni will elevate the first two ribs. When the ribs are fixed the movements of the neck will vary with the action of the different muscles. If the two posterior scaleni of one side contract, the neck will be inclined laterally towards the muscles acting; but if those of both sides come into play at once—the one set antagonizing the other—the vertebral column will remain upright. Should the anterior scaleni of both sides act the neck would be bent forwards, in consequence of their attachment in front of the spine.

Another muscle, the *serratus magnus*, lies in the lower and outer angle of the triangular space, viz., where the omo-hyoideus and the trapezius meet; it is concealed by the trapezius.

POSTERIOR TRIANGULAR SPACE OF THE NECK.

The intermuscular interval on the side of the neck, named the posterior triangular space, is narrow before the fascia is removed, like the corresponding hollows opposite the joints, but in the Drawing the space is delineated as it appears after dissection. The great artery of the upper limb with some smaller branches, and the cervical nerves, together with much fat and interspersed lymphatic glands, are contained in this hollow.

This interval reaches from the clavicle to the back of the head. It is bounded in front by the sterno-mastoideus, B, and behind by the trapezius muscle, D. By its dissection greater apparent length is given to the neck, in consequence of the teguments being removed from part of the head.

Narrower above than below, the space is said to be triangular. Rather it is flask-shaped, with the small part directed upwards. As low as the letter D, the hollow is shallow, and the sides nearly straight; but beyond that spot it becomes deeper, and is widened in consequenae of the posterior border being curved. When in its natural position, the upper part of the sterno-mastoid projects farther back towards the trapezius than is indicated in the Drawing.

Stretched over the space are the skin, the subcutaneous fatty layer containing the platysma, A, and the deep cervical fascia. And the floor of the hollow is formed by the superficial stratum of the muscles of the side of the neck, in the following order. Beginning above, the splenius capitis, C, is first met with; and below it lies the levator anguli scapulæ, divided into parts and marked by E, E. Farther down comes the scalenus medius, F; and near the clavicle the serratus magnus projects above the first rib, but this would be visible under the trapezius only in a front-view.

The space is divided into two unequally-sized parts by the small omo-hyoideus muscle, H—the lower being designated clavicular, and the upper occipital.

The *occipital part*, much the larger of the two, occupies nearly the whole length of the neck. It has the same bounds in front and behind as the large hollow; and it is limited below by the omo-hyoideus, H. Its depth increases towards the lower boundary, and in it are contained chiefly nerves, with some small vessels, and lymphatics.

The *nerves* issue from beneath the sterno-mastoid muscle, and unite in a plexiform manner—the upper nerves entering the cervical, and the and lower the brachial plexus.

From the nerves, 1, and 2, of the cervical plexus, superficial branches are directed upwards and downwards:—The ascending set reach the fore part of the neck, the ear and contiguous part of the face, and the back of the head; and the descending set, more numerous than the other, are continued through the space to the integuments of the top of the chest and shoulder.

The lower cervical nerves join in the brachial plexus, 11. These trunks are inclined downwards through the lower end of the occipital part, and through the clavicular part of the triangular space to the axilla: they give few branches, and their position will be referred to again.

One large nerve, 13, the *spinal accessory* (eleventh cranial nerve), crosses the space obliquely downwards and backwards, from the border of the sterno-mastoideus to the under surface of the trapezius.

Vessels. The arterial branches are small in size, and supply the surrounding muscles: they appear behind the sterno-mastoideus. The lowest and largest is the transverse cervical artery, c, which passes beneath the trapezius. Veins accompany the arteries, their size corresponding with that of their companions.

The *clavicular part* of the posterior triangular space has its side formed by the clavicle, L, and the omo-hyoideus, H; and its base or fore part by the sterno-mastoideus, B. Towards the surface it is covered by the same layers as the great triangle; and the floor is constructed by the scaleni muscles, the serratus magnus, and the first rib.

Larger before than behind, it is placed nearly opposite the middle third of the clavicle. It is about one inch and a half long, and an inch wide in front after the dissection; but until the omo-hyoideus has been displaced, this muscle will lie closer to the clavicle, diminishing thus the width. Contained in it are the subclavian artery, a, the brachial plexus, 11, and the external jugular vein, k, with their offsets, together with lymphatics and the usual fat.

Arteries. The subclavian trunk, a, crosses the space from within out. In front it issues from beneath the anterior scalenus, G; and it disappears below beneath the clavicle. Along the side of the space formed by the clavicle, the supra-scapular vessels, b, lie under cover of that bone. And at the corner where the omo-hyoideus meets the sterno-mastoideus, the transverse cervical vessels, c, cross the hollow.

Veins. If the subclavian vein is full it may appear beneath the clavicle, though it lies usually at a lower level than the artery. The external jugular vein, k, is directed across the space, to join the subclavian vein below: companion veins, l and n, of the transverse cervical and supra-scapular arteries enter it near the clavicle.

Nerves. External to the artery, or higher in the neck than it, the large bundles of nerves entering the brachial plexus are directed downwards in their course to the arm-pit: they have a deep position like the artery, and occupy the interval between the vessel and the omo-hyoid muscle. Near the outer part of the space they approach closer to the vessel, and serve as a valuable guide to it from the constancy of their position, and their white appearance and firm feel. Over the space descend the superficial branches of the cervical plexus: these must be divided in an incision into the neck.

The size of the clavicular part of the triangular space varies much with the condition of the bounding muscles. Alterations in length will be determined by the attachment of the trapezius and sterno-mastoideus to the clavicle, for if one or both should reach farther than usual on that bone, the intermuscular space must be diminished accordingly. The width will be dependent upon the size and the situation of the omo-hyoideus, H. When the muscle is wide, or lies close to the clavicle, the dimensions from above down of the clavicular part of the triangular space will be less than when the muscle is narrow, or is placed at a greater distance from the bone. In some bodies the omo-hyoideus arises from the back of the clavicle, and conceals the subclavian artery, so that there is not any interval in the usual place between the muscle and the collar-bone.

Differences in depth will arise from varying states of the neighboring parts. In a long and thin neck, with low and flat clavicles, the depth is not so great as in a short and thick neck with prominent and much curved collar-bones. Changes in the position of the shoulder will give rise also to variations in depth. Thus if the shoulder is depressed by drawing down the arm, the space is as shallow as it can be made; whilst raising the shoulder gives to the hollow its greatest depth. And by forcing upwards the shoulder the clavicle can be carried as high as, or even higher than the level of the omo-hyoid muscle and the subclavian artery.

ARTERIES IN THE TRIANGULAR SPACE.

In the lower part of the triangular space are contained the trunk of the subclavian artery, and some of its branches. Towards the ear are other small arteries, which are derived from the external carotid trunk.

- *a.* Subclavian artery.
- *b.* Supra-scapular artery.
- *c.* Transverse cervical artery.
- *d.* Cutaneous branch of the subclavian.
- *e, f.* Branches of the ascending cervical artery.
- *g.* Posterior auricular artery.
- *h.* Cutaneous offset of the posterior auricular.

Subclavian artery, a. The third part of the arch of the subclavian trunk (Plate xviii.) lies in the clavicular portion of the posterior triangular space; and it has the following anatomy.

Its extent is marked by the outer edge of the anterior scalenus, G, on the one side, and the lower border of the first rib on the other (below the clavicle). The vessel is directed outwards at first, about an inch above the clavicle, and it passes downwards finally under the most prominent point of that bone. Superficial to the artery are the common coverings of the space, viz., the skin, the cutaneous fat with the platysma, and the deep fascia; and as it is about to pass under the clavicle and the subclavius the supra-scapular artery and vein cross in front. Underneath the vessel lie the middle scalenus, F, and the first rib.

Its companion vein, subclavian, is arched like the artery (Plate xviii.), but it is placed lower in the neck, and beneath the clavicle. Crossing the artery near the scalenus is the external jugular vein, k, whose branches may form a plexus over it.

The nerves of the brachial plexus, 11, lie above the artery near the scalenus anticus, and gradually approach it below, so that, at the clavicle, the trunk formed by the last cervical and first dorsal touches the vessel, and may be mistaken for it in the operation of tying the subclavian. A small branch, 10, to the subclavius muscle is directed across the artery. Superficial to the clavicular space are the descending cutaneous branches of the cervical plexus, which will be cut in the operation for ligature.

Offsets of the artery. As a rule this part of the subclavian trunk does not furnish any named branch. A *cutaneous* offset, d, took origin from the vessel in this body, but it springs commonly from the supra-scapular artery, b, near the external jugular vein.

Compression of the artery. As the subclavian artery is uncovered by muscle whilst it crosses the triangular space, it may be compressed at the lower part of the neck during life. Its position is marked on the surface by the most prominent part of the clavicle; and if the thumb is pressed firmly downwards and backwards behind that point of the bone towards the first rib, the circulation in the vessel may be stopped. Sometimes the top of a key padded may be used more advantageously than the thumb.

Ligature of the third part of the subclavian artery is practised commonly for aneurism of the axillary trunk; and as this operation may be rendered more difficult by the unusual position of the subclavian vessels, and by unusual states of the surrounding parts, the conditions complicating it will be first reviewed.*

---

* The summary here made has been derived from the facts made known by Mr. Quain's researches on the Surgical Anatomy of the Arteries, in the work before quoted.

Alterations affecting the artery. Commonly the arch of the vessel rises about an inch above the clavicle (Quain), but it may be lowered to the level of, or sink beneath the bone; and on the other hand it may be elevated as high as one inch and a half above the collar-bone. Occasionally the artery passes over or through the anterior scalenus, instead of beneath it. When the artery has either the higher level, or the more superficial position, it will be rendered less deep, and will be more easy to find in an operation.

One or two branches for the shoulder, viz., posterior scapular and supra-scapular, may spring from this part of the artery. If such branch or branches should be seen in an operation, greater security against secondary hæmorrhage would be obtained by tying one or both, than by leaving either free to convey blood into or from the trunk near the ligature.

Alterations in the surrounding parts. With a thin and long neck and a flat clavicle, there is a prospect of a less tedious operation than in the opposite states of those parts, because the artery will be nearer the surface.

Muscular fibres may cover the artery as before said, p. 126, the clavicular attachments of the sterno-mastoid and the trapezius being lengthened, or the omo-hyoid arising from the clavicle. Also in axillary aneurism high in the arm-pit the clavicle may be carried upwards considerably above the level of the subclavian artery. Under these circumstances the operation of ligature would be made more difficult, as the artery must be sought behind the raised bone in the one case, and beneath the muscular fibres in the other.

The subclavian vein rises sometimes as high as the level of the clavicle; and it has been found twice beneath the anterior scalenus with the subclavian artery; both changes in its position would cause it to be more endangered in the steps of an operation. The external jugular may be moved outwards from the sterno-mastoideus as far as the middle of the clavicle, so that its trunk and branches would lie in the centre of an incision to reach the artery: this position of the vein may so interfere with the access to the artery as to render expedient division of the vein, and ligature of the ends.

Steps of the operation of ligature. Taking the most prominent part of the clavicle as the superficial guide to the position of the artery, draw down the loose skin of the neck, and cut for two inches and a half along the clavicle—the line of the vessel marking the centre of the cut—so as to divide on the bone the skin, the fat and the platysma, and the superficial nerves and vessels.* Let this cut be next moved rather above the clavicle; and let the operator divide the deep fascia, and find his way vertically downwards to the artery, looking out for the intermuscular interval between the trapezius and sterno-mastoideus, and for that between the omo-hyoideus and clavicle, and incising any muscular fibres which interfere with his progress. After the muscles have been passed the surgeon proceeds cautiously, not letting the knife pass beneath the clavicle to wound the supra-scapular vessels or the subclavian vein, and using at this stage the outer, rather than the inner part of the wound. Towards the inner end of the incision the external jugular vein with branches will soon be met with; and it may be either drawn inwards, or divided and tied, according to the impediment it offers to reaching the artery.

To find the artery in the bottom of the wound, look to the outer end for the firm and white cords of the brachial plexus, which serve as the deep guide; and when these are recognized the artery will be found lower down, i. e., between them and the first rib.† After the artery has been laid bare by the removal of some fat and a slight sheath, the aneurism needle should be entered in the outer angle of the wound, where the handle can be depressed so as to make the point with the thread turn under the vessel.

Arterial branches. The smaller arteries laid bare in the dissection are derived from two arterial trunks. Behind the sterno-mastoideus they are offsets of the subclavian or limb artery; and the branches in front of the muscle, or piercing it (except the lowest), spring from the carotid or neck artery.

The *supra-scapular artery, b,* comes from the first part of the subclavian trunk, and runs behind the clavicle with its vein to the upper border of the scapula: it ends on the dorsum of that bone.

---

* If such a superficial vessel as that marked *d*, in the Drawing, should arise from the subclavian trunk, division of it at this stage would be followed by considerable hæmorrhage, and ligature of it would probably be needed before the operation could be continued.

† The projection or tubercle on the first rib, at the attachment of the anterior scalenus muscle, is said by some authors to serve as the deep guide to the vessel, but this eminence is seldom prominent enough to be felt by the finger.

An offset from the supra-scapular to the integuments arises near the sterno-mastoid: in this instance it comes from the third part of the subclavian, and is marked *d*.

Transverse cervical artery, c. It arises in common with the preceding, and crossing the side of the neck above the arch of the subclavian artery, courses beneath the trapezius: here it furnishes a large branch (superficial cervical), and bends finally along the base of the scapula with the name *posterior scapular,* and supplies the muscles inserted into the vertebral border of that bone.

In the posterior triangle it gives many branches to the levator anguli scapulæ, and to the lymphatic glands and the fat.

Two small arteries, *e* and *f,* are offsets of the *ascending cervical artery* (a branch of the subclavian): they are distributed to the muscles on the side of the neck, and to the areolar tissue and the glands of the triangular space.

The *posterior auricular artery, g,* issues in front of the sterno-mastoideus, and ascends to the back of the ear and the contiguous part of the head.

A cutaneous offsets, *h,* courses over the sterno-mastoid muscle, and accompanies the small occipital nerve.

Perforating branches. After piercing the sterno-mastoid muscle these small arteries supply the platysma and the teguments.

SUPERFICIAL VEINS OF THE NECK.

In the neck there are two superficial or jugular veins, a lateral and an anterior. Only the lateral vein and its branches appear in the Drawing: the other is figured in Plate xvi.

k. External jugular vein.
l. Transverse cervical vein.
n. Supra-scapular vein.
o. A subcutaneous vein.

The *external jugular vein, k,* conveys blood from the head to the subclavian vein, and gathers blood also from the superficial parts of the neck. It begins in the parotid gland by the union of the temporal and internal maxillary veins (Plate xvii.); and becoming superficial, it descends beneath the platysma muscle, A, to the lower part of the neck, where it sinks through the fascia and ends in the subclavian vein (Plate xviii.). Its common position in the neck would be marked by a line from the angle of the jaw to the middle of the clavicle, though in the Plate it is placed internal to that line.

At the upper part of the neck the vein is small in size, receiving only few branches, but for an inch and a half at the lower end, it is dilated behind the sterno-mastoid muscle: here it receives veins from the shoulder, viz., the transverse cervical, *l*, the supra-scapular, *n*, and some cutaneous veins—one being marked with *o*. A pair of valves exists both above and below the lower dilatation. The lower pair is close to the clavicle, and is less complete than the other, for it allows blood to pass in a reflex course from the subclavian vein. The upper pair is found just after the vein crosses the sterno-mastoid muscle, and acts perfectly, as it permits the blood to flow only in one direction—from above down.

Bloodletting in the external jugular vein, is seldom had recourse to now, but the steps of the operation are the following:- The downward current of the blood is stopped by pressure of the thumb near the clavicle. A cut is then made obliquely upwards and backwards across the vein, to incise the vessel and the fibres of the platysma to the necessary extent. As long as the pressure on the vein remains the blood issues through the opening, but when the thumb is removed the flow stops, because the blood finds its way by the usual channel into the subclavian. After the operation is finished the wound is to be closed by adhesive plaster.

Under some conditions air may enter the vein during the operation of bloodletting. As long as the blood runs freely, and the breathing is regular, the accident is not likely to happen: but if the breathing becomes labored, or if the opening is not closed as soon as the flow of blood stops, air may be drawn into the vein.

In suspended animation the external jugular is sometimes opened with the view of relieving the over-distended right side of the heart;[*] and this practice is founded on the fact that blood will enter the jugular below from the subclavian vein. At the same time the blood can flow

---

[*] A more general employment of this practice is recommended by Dr. Struthers, in a paper "On Jugular Venesection in Asphyxia." Edin. Med. Journal for November, 1856.

downwards through the anterior jugular in the usual way (Plate xviii.), so as to relieve simultaneously the congested heart and head.

NERVES IN THE POSTERIOR TRIANGULAR SPACE.

Parts of the cervical and brachial plexuses of nerves, with one cranial nerve—the spinal accessory, are included in the dissection.

1. Third cervical nerve.
2. Fourth cervical nerve.
3. Great auricular nerve.
4. Small occipital nerve.
5. Superficial cervical nerve.
6. Superficial descending branches of the cervical plexus.
7. Nerve to the rhomboideus.
8. Nerve to the serratus magnus.
9. Branches to the trapezius.
10. Nerve to the subclavius.
11. Upper part of the brachial plexus.
12. Supra-scapular nerve.
13. Spinal accessory nerve.
14. Posterior auricular nerve.
† Nerve to the levator anguli scapulæ from the cervical plexus.

The *cervical plexus* is formed by the union of the upper four cervical nerves; and it lies beneath the sterno-mastoideus, B, and on the levator anguli scapulæ, E. Only the lower part of the plexus comes into the posterior triangular space, and from it spring muscular, and ascending and descending tegumentary branches.

Ascending branches. These consist of the three following nerves, which are directed to the ear, the occiput, and the fore part of the neck.

The *great auricular nerve*, 3, courses near the external jugular vein to the lobe of the ear, and ends in the integuments of the hinder and outer parts of the pinna. One offset joins the posterior auricular nerve, 14, and others are directed forwards to the integuments over the parotid gland: some long slender branches pass through the parotid to join the facial nerve (Plate xvii.).

The *small occipital nerve*, 4, lies along the posterior border of the sterno-mastoideus, and perforating the deep cervical fascia near the head, ramifies in the scalp of the occipital region.

The *superficial cervical nerve*, 5, is often represented by several small nerves, as in the Drawing, and is therefore very variable in its size: it is distributed to the platysma, and to the integuments of the neck in front of the sterno-mastoid muscle.

Descending branches. The chief of these, two or three in number, belong to the teguments of the shoulder and the upper part of the thorax; but some offsets are directed backwards to the integuments over the trapezius muscle, from the clavicle nearly to the head.

A large nerve, 6, divides into three:—one crosses the attachment of the sterno-mastoideus to the clavicle, another lies over the insertion of the trapezius into the same bone, and the third crosses the middle of the clavicle; they extend two or three inches below the collar bone, the inner nerves reaching least far.

Muscular offsets. Only a few of these are now visible. One, †, enters the levator anguli scapulæ. Others, 9, pass beneath the trapezius supplying it; and they join beneath that muscle with the spinal accessory nerve, 13.

Brachial plexus.—The lower four cervical nerves, and the first dorsal nerve (in part), give rise to the large bundles of nerves marked, 11; but in the side view presented to the Artist the arrangement of the several nerves entering the plexus could not be shown as in Plate xviii.

The plexus extends under the clavicle to the axilla, where it terminates in nerves for the upper limb; and all the muscular offsets in the neck come from the fifth and sixth cervical nerves, with the exception of small branches to the longus colli and the scaleni.

The *nerve to the rhomboid muscle,* 7, pierces the fibres of the scalenus medius, and is inclined backwards beneath the elevator of the angle of the scapula.

Nerve to the serratus magnus, 8 (posterior thoracic). This nerve issues through the scalenus medius, below the preceding, and is continued ber. ath the cords of the plexus to the axilla. See Plate ii., 5.

The *nerve to the subclavius,* 10, passes in front of the subclavian artery to the under surface of its muscle.

The *supra-scapular nerve,* 12, accompanies the omo-hyoid muscle to the back of the scapula, and supplies the supra and infra-spinate muscles, the shoulder-joint, and the blade bone.

The two remaining nerves, which are seen in this part of the neck, belong to the cranial set.

The *spinal accessory nerve,* 13 (eleventh cranial), pierces the sterno-mastoideus, and ends in the trapezius, after crossing the posterior triangular space, where it joins the spinal nerves. Under the trapezius it communicates with the nerve marked, 9, before it enters the fleshy mass.

The *posterior auricular,* 14, a branch of the facial or seventh cranial

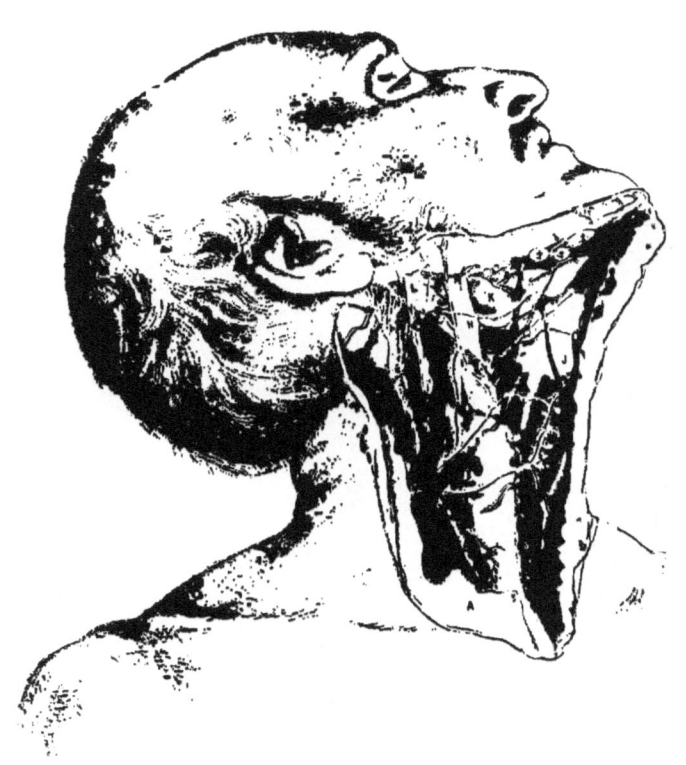

nerve, ascends in front of the mastoid process, and being joined by the great auricular nerve, splits into two:—one piece belongs to the integuments of the back of the ear, and the retrahent muscle of the pinna; and the other supplies the hinder belly of the occipito-frontalis muscle, and the integument contiguous to it.

Lymphatics. Beneath the fascia lies a collection of cervical lymphatic glands in the clavicular part of the posterior triangle. They communicate below with the lymphatics of the axilla; and above with those about the ear and the occiput, by means of the superficial lymphatic vessels and glands accompanying the external jugular vein. Beneath the sterno-mastoideus they join also the deep glands by the side of the carotid vessels.

DESCRIPTION OF PLATE XVI.

THIS Plate exhibits a surface-view of the side of the neck, in front of a line from the mastoid process to the inner end of the clavicle.

Supposing the skin thrown aside, as in the Figure, the thin fleshy fibres of the platysma will appear through a slight fatty covering, and may be readily cleaned. This muscle may be then raised towards the jaw by a cut over the sterno-mastoideus, the superficial veins and nerves being traced out at the same time. Before the removal of the deep fascia the subjacent muscles should be fixed in their natural position by stitches, to prevent their slipping out of place when the investing sheaths are taken away.

Afterwards the areolar tissue and fat are to be cleared out between the jaw and the hyoid bone, and from the whole surface of the space laid bare.

SURFACE VIEW OF THE FRONT OF THE NECK.

The prominent sterno-mastoid muscle, B, divides into two the side of the neck; and in front of it is a slight hollow, which is most marked near the jaw, and is wider above than below.

In front of the sterno-mastoideus lie the elevator and depressor muscles of the hyoid bone,—the former extending downwards from the lower jaw, and the latter reaching upwards from the chest and shoulder.

Below the side of the jaw is the submaxillary gland, K, with a chain of small lymphatic glands reaching backwards to the sterno-mastoid muscle; and a lymphatic gland, with a small artery entering it, is lodged just above the body of the hyoid bone. Between the jaw and the ear the parotid gland, L, is wedged in.

No large arterial trunk can be seen on the surface of the neck as long as the sterno-mastoideus keeps its natural position;* and this Plate teaches also that no triangular space containing the large cervical bloodvessels is observable until that muscle has been displaced, as in Plate xviii.

A few small arteries reach the surface. Thus, the facial artery, a, with its vein winds over the submaxillary gland and the jaw in front of the masseter muscle, and gives forwards the submental branch, b, below the jaw; whilst opposite the back of the hyoid bone the lingual vessels, c, appear for a short distance. Issuing from beneath the sterno-mastoideus are small cutaneous offsets, e, of the upper thyroid artery—one, d, entering the superficial lymphatic gland near the hyoid bone; and piercing the sterno-mastoideus are other cutaneous arteries, f, of the subclavian and external carotid trunks. Near the ear a cutaneous branch, g, of the posterior auricular artery, escaping beneath the parotid gland, crosses over the sterno-mastoideus.

Two superficial jugular veins are directed from above down through the anterior part of the neck. One, the external jugular, h, crosses the sterno-mastoideus from before back; and the other, the anterior jugular, l, lies in front of that muscle, and near the middle line of the neck.

Cutaneous nerves cross from behind forwards, spreading out over the region dissected. The nerve marked, 1, is the cervical part of the seventh cranial nerve, which reaches as low as the hyoid bone; and the nerve, 2, is a branch of the cervical plexus to the teguments below the preceding.

---

* Anatomists depict and describe the common carotid artery as uncovered by the sterno-mastoideus at its upper end. And the directions of surgeons for placing a ligature on that bloodvessel are based on the same inaccuracy.

'ing by

MUSCLES AND THE CERVICAL FASCIA.

Most of the muscles laid bare will be described more fully in the explanation of the following Plate; but as the natural state of the sterno-mastoideus, and its connection with the cervical fascia would be destroyed by the deeper dissection, these will be noticed below.

A. Platysma myoides, cut.
B. Sterno-cleido-mastoideus.
C. Thyro-hyoideus.
D. Omo-hyoideus.
E. Sterno-hyoideus.
F. Anterior belly of the digastric muscle.
H. Stylo-hyoideus.

J. Hyoid bone.
K. Submaxillary gland.
L. Parotid gland.
N. Process of the deep cervical fascia fixing the sterno-mastoideus.
† Lymphatic glands.

The *sterno-cleido-mastoid* muscle, B, incases somewhat the narrowed part of the neck by the elongation of its edges forwards and backwards. The anterior curved border is manifest in the Drawing, and it is kept in this position by a piece of fascia, N, which is attached to the lower jaw.

The muscle covers the carotid bloodvessels as high as the digastricus, and even when the head is thrown backwards.

In the operation of tying the common carotid artery the muscle would have to be dissected back for some distance before the line of the vessel is reached; and pressure on the artery must be made through the muscle. This fleshy covering gives protection to the large vessels; and these cannot be injured in wounds of the neck unless the muscle is cut.

Deep cervical fascia. The special fascia of the neck invests the muscles with sheaths. Most of it has been removed in cleaning the muscles; but a strong process marked, N, has been left for the purpose of showing its connection on the one hand with the sterno-mastoideus, and on the other with the angle of the lower jaw. The office of this piece is to keep curved the anterior border of the sterno-mastoideus, for as soon as it is cut the edge takes a straight direction, as in Plate xvii.

cles of
jaw

CONNECTIONS OF THE SALIVARY GLANDS.

On each side there are three salivary glands in contact with the lower jaw. One is lodged behind the ramus and angle of the bone, and is named parotid: another is partly covered by the side of the jaw—the submaxillary; and the third, the sublingual, lies beneath the front of the tongue.

The *parotid* is the largest of the salivary glands. It is placed between the jaw in front, and the ear with the mastoid process and the sterno-mastoideus behind; and it projects downwards beyond the level of the jaw, where the process, N, of the cervical fascia separates it from the submaxillary gland.

Towards the surface the gland is flat, and is covered by the deep cervical fascia: on it rest one or more lymphatic glands. Its deep part is very irregular in form, and sends downwards prolongations around the styloid process.

Several vessels and nerves pass through the substance of the parotid, and the position of these may be studied in Plate xvii. The external carotid artery, b, ascends through the gland giving off the auricular, temporal, and internal maxillary branches. The external jugular vein, r, begins by the union of the temporal and internal maxillary branches, and passes downwards superficially to the carotid. The facial nerve, 4, traverses the gland from behind forwards, over the artery, and is joined by offsets of the great auricular nerve. Close to the ear the cutaneous part, 11, of the auriculo-temporal nerve is directed upwards by the side of the temporal artery.

The excretory duct of the gland (ductus Stenonis) leaves the fore part, and piercing the buccinator muscle, opens into the mouth opposite the second molar tooth of the upper jaw (Plate xx.).

In enlargement of this gland the swelling will project downwards at first towards the deep vessels and nerves in front of the spine, and then into the neck along the sterno-mastoideus; but extension towards the surface will be delayed by the strong fascia binding it down. Much pain will attend the swelling of the glands in " mumps " and other affections, just as in all inflamed glandular parts that are prevented expanding by the firmness of the encasing sheaths.

The swelling and abscesses in front of the ear in scrofulous children are occasioned by inflammation of the lymphatic glands on the surface of the parotid.

The *submaxillary gland*, K, is not surrounded by such unyielding structures as the parotid; for, though concealed somewhat by the side of the maxilla, it projects down the neck for an inch or more in front of the angle of the jaw. Superficial to it are the integuments and the platysma with the deep fascia; and beneath it is the mylo-hyoid muscle. In front it is bounded by the anterior belly of the digastric, F; below by the digastric and the stylo-hyoid, H; and behind by the process, N, of the deep cervical fascia which intervenes between it and the parotid. Over the surface wind the facial vessels, a.

The glands consists of larger lobules than the parotid; and from its deeper surface the duct (Wharton's) is continued to the floor of the mouth: the course of the duct is evident in Plate xxii. of the submaxillary region.

The *sublingual gland* projects in the floor of the mouth under the front of the tongue, where it forms a lengthened swelling. Placed deeply under the side of the jaw, close to the symphysis, its connections will be indicated in Plate xxii.

The *lymphatic glands* marked thus, †, are three or four in number, and lie along the base of the jaw, superficial to the submaxillary gland: they receive vessels from the submental artery, b. In scrofulous children these glands may enlarge, and suppurate.

In the middle line, just above the hyoid bone, is a small lymphatic gland, which receives a vessel, d, from the upper thyroid artery.

SUPERFICIAL ARTERIES OF THE NECK.

In comparison with the superficial veins the arteries appearing on the surface are few, and are small in size. None except the facial, a, and the lingual, c, are large enough to furnish serious hæmorrhage in superficial wounds; but in cuts involving the muscles, the large trunks displayed in Plate xvii. may be opened.

a. Facial artery and vein.
b. Submental branch of the facial.
c. Lingual artery and veins.
d. Offset of the upper thyroid artery to a lymphatic gland.

ee. Superficial offsets of the upper thyroid artery.
ff. Branches of the subclavian and carotid trunks perforating the sterno-mastoideus.
g. Cutaneous offset of the posterior auricular.

The anatomy of the several arteries will be given with the description of Plate xvii.

SUPERFICIAL VEINS OF THE NECK.

Two superficial jugular veins, and the facial and linguals veins, appear in this dissection.

h. External jugular vein.
l. Anterior jugular vein.

n. Facial vein.

The *external jugular* vein, h, is figured in the part of its course which is superficial to the deep fascia of the neck, and is concealed by the platysma muscle, A (p. 131).

The *anterior jugular* vein, l, begins in the teguments below the chin, and communicates with a branch of the facial vein. Lying superficially near the anterior edge of the sterno-mastoideus, it sinks through the cervical fascia near the sternum, and opens into the subclavian vein (Plate xviii.). It unites commonly by a branch with the external jugular.

CUTANEOUS NERVES OF THE FRONT OF THE NECK.

The facial nerve and branches of the cervical plexus supply the superficial structures of the neck.

1. Cervical part of the facial nerve.
2. Superficial cervical nerve.

3. Great auricular nerve.

The *infra-maxillary* branch, 1, of the facial nerve, issuing from beneath the parotid, sends forwards offsets beneath the platysma as low as the hyoid bone: it supplies that muscle and the integuments.

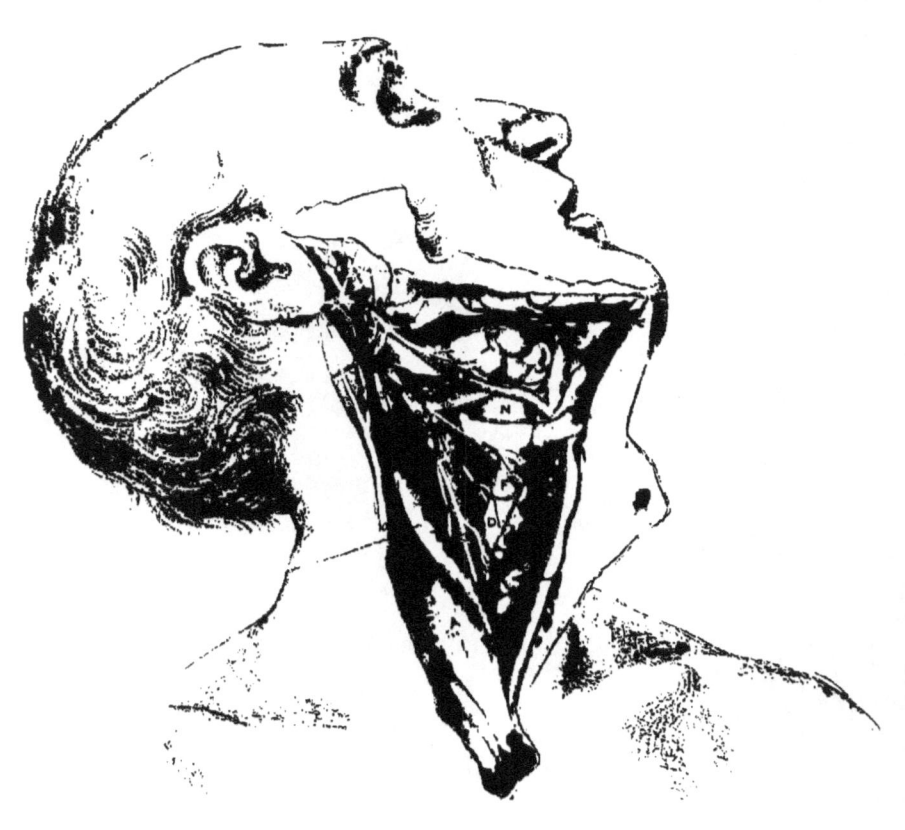

The *superficial cervical nerve*, 2, is bent forwards under the platysma, and its branches pierce the muscle to supply the integuments between the hyoid bone and the sternum. Above, it joins the facial nerve: and it is said to give offsets to the lower part of the platysma.

The *great auricular nerve*, 3, ascends by the side of the external jugular vein to the ear, and ends as before said (p. 133).

DESCRIPTION OF PLATE XVII.

The anterior triangular space of the neck, as it appears after displacement of the sterno-mastoid muscle, is shown in this Figure.

Whilst the skin is being reflected forwards, the platysma muscle may be denuded; and this muscle and the deep cervical fascia should be then raised. Afterwards the parotid gland is to be picked out of the hollow between the ear and the jaw, and the areolar tissue and the fat are to be removed from the space, as is seen in the Drawing, without injury to the numerous vessels and nerves.

ANTERIOR TRIANGULAR SPACE.

This intermuscular space corresponds with the surface-depression between the jaw and the sternum, and contains the carotid bloodvessels with their companion veins and nerves, and some salivary and lymphatic glands.

Triangular in form, with the base upwards, it is bounded in the following way. Behind is the reflected sterno-mastoideus with the ear; and in front the space reaches to the middle line of the neck. At the base lies a jaw-bone; and the apex touches the top of the sternum.

Stretched over the hollow are the teguments, with the platysma muscle and the deep cervical fascia; and in the floor the air and food passages are lodged, covered by the muscles of deglutition.

The depth increases from below upwards; and it is greatest along the upper two thirds of the sterno-mastoideus and the front of the ear, where the great vessels and nerves are placed, but it diminishes gradually towards the front.

Along the middle of the neck lie certain well-marked prominences, which can be felt readily by the finger during life, and serve as guides in in operations on the vessels and the windpipe. About two inches from the lower jaw, when this is raised, projects the narrow firm line of the hyoid bone, which is marked J, in the preceding Plate. A finger's breadth below that bone the prominence of the thyroid cartilage of the larynx (pomum Adami) is met with. Still lower, about an half an inch, comes the firm cricoid or ring cartilage of the larynx,—a prominence less than the former; and between the two is a slight hollow, opposite the crico-thyroid membrane, through which the knife is sunk in the operation of laryngotomy. From this point to the sternum the tube of the windpipe and the thyroid body carry forwards the muscles: the former can be recognized by the finger.

Behind the os hyoides and the larynx and trachea lies the pharynx with the œsophagus.

Position of arteries. Opposite the level of the cricoid cartilage the large trunk of the common carotid, *a*, escapes from beneath the depressor muscles of the hyoid bone. In company with the internal jugular vein it lies between the pharynx and the spine, and ascends under cover of the sterno-mastoideus, A, to the upper border of the thyroid cartilage, where it splits into the external carotid, *b*, and internal carotid, *c*. From the point of division these two arteries are continued in the direction of the parent trunk to the interval between the ear and the jaw, and they end in the following way:—one (external) is consumed in offsets outside the cavity of the skull; and the other (internal) is distributed chiefly to the brain, without furnishing branches to the neck. Neither vessel is visible till after the sterno-mastoideus has been displaced. See Plate xvi.

Position of veins. By the side of the common carotid artery is the internal jugular vein, *p;* and it is continued to the base of the skull along the internal carotid trunk. In the upper narrowed part of the space between the jaw and the ear the external jugular vein begins; but it then runs downwards over the sterno-mastoideus. Near the middle line of the neck the anterior jugular vein, *s*, descends (Plate xvi.); and it passes beneath the sterno-mastoideus at the lower part of the neck.

Position of nerves. Many nerve trunks lie in contact with the great bloodvessels, and most of them accompany those vessels to the base of the skull.

Superfical to the sheath of the vessels where the common carotid may be tied, is the descendens noni nerve, 3. A little above the hyoid bone the hypoglossal nerve, 14, is directed forwards over both carotid arteries; and in front of the ear the branches of the facial nerve, 4, cross over the external carotid.

On a line with the base of the jaw-bone the glosso-pharyngeal nerve is inclined inwards between both arteries.

In the sheath, between and parallel with the vein and artery, the vagus nerve extends through the neck (Plate xxiv.); two of its branches, the superior laryngeal, 1, and the external laryngeal, 2, being directed inwards to the larynx.

Beneath the sheath the cord of the sympathetic nerve (Plate xxiv.) rests on the spinal column.

External or posterior to the sheath for a short distance is the spinal accessory nerve, 13, as this issues from beneath the digastric muscle, R.

One small nerve is altogether removed from the sheath: it is the mylo-hyoid branch, 12, of the inferior maxillary nerve (Plate xxi.), and escapes from under the jaw-bone.

Glands of the space. Two large salivary glands, the parotid and sub-maxillary, which are seen in Plate xvi., where they are marked L and R, occupy the base of the triangular space.

The lymphatic glands have been cleared away in the dissection: one set lies along the jaw-bone (Plate xvi.); and the other (deep cervical) is placed along the side of the jugular vein, under the sterno-mastoideus.

MUSCLES OF THE FRONT OF THE NECK.

The muscles occupying the upper and fore parts of the triangular space converge to the os hyoides—the upper set elevating, and the lower set depressing that bone.

A. Sterno-mastoideus.
B. Stylo-hyoideus.
C. Omo-hyoideus—anterior belly.
D. Sterno-thyroideus.
F. Thyro-hyoideus.

H. Digastricus—anterior belly.
N. Hyo-glossus.
P. Stylo-hyoideus.
R. Digastricus—posterior belly.
S. Masseter.

Depressors of the os hyoides. These muscles cover the trachea and larynx, and are marked B, C, and D.

Omo-hyoideus, C. The anterior belly of this muscle crosses the common carotid artery and jugular vein just below the cricoid cartilage, and is inserted into the body of the hyoid bone close to the great cornu. For the posterior belly, see Plate xv. and page 123.

Sterno-hyoideus, B. The muscle arises from the inner surface of the sternum and first rib, and is inserted into the middle of the body of the hyoid bone.

Sterno-thyroideus, D, arises lower in the chest than the preceding, though like it from the sternum and the rib, and is inserted into the oblique line on the thyroid cartilage, where it is continuous with the following.

The small *thyro-hyoideus,* F, joining the preceding below, is inserted into the anterior half of the great cornu, and into part of the body of the os hyoides.

This group of muscles is covered partly by the sterno-mastoideus; and it conceals the windpipe and the thyroid body, and the sheath of the great bloodvessels. An interval separates the muscles of opposite sides along the middle line of the neck, except for about an inch above the sternum, where the sterno-thyroid muscles touch.

Action. Commonly the muscles act from the sternum, and draw down rapidly the os hyoides as soon as the morsel of food or the fluid to be swallowed has passed the upper aperture of the larynx. If they take their fixed point above, the sterno-hyoid and the sterno-thyroid will assist in dilating the chest in laborious breathing; and the small thyro-hyoid, F, will raise and tilt backwards the thyroid cartilage—relaxing thereby the vocal cords.

Elevators of the hyoid bone. These muscles are more numerous than their antagonists, for some extrinsic muscles of the tongue help to raise the os hyoides: the deeper muscles of the set may be referred to in Plate xxii.

Stylo-hyoideus, P. Arising near the root of the styloid process, the muscle is divided into two parts, between which passes the tendon of the posterior belly, R, of the digastricus; and it is inserted into the body or the great cornu of the os hyoides, joining the aponeurosis of the digastricus.

The *digastric* muscle consists of two fleshy parts with an intermediate tendon.

The posterior belly, R, is fixed to the groove beneath the mastoid process of the temporal bone; and the anterior, H, is attached to the jaw close to the symphysis. Below, the muscle is connected to the surface of the body of the hyoid bone by a thin aponeurosis, which joins the anterior belly and the fore part of the tendon.*

The digastric incloses with the jaw a space in which the two superficial salivary glands are lodged. And the posterior belly marks the spot at which the carotid bloodvessels become deep and inaccessible: the position of this part of the muscle corresponds with a line on the surface from the mastoid process to a point half an inch above the hyoid bone.

The *mylo-hyoid* muscle descends from the jaw-bone to the body of the os hyoides: it lies beneath the anterior belly of the digastric, and in Plate xxii., where it is reflected, it may be seen to join its fellow along the middle line of the neck.

The *genio-hyoideus* is beneath the preceding. Plate xxii. shows it in position, reaching from the jaw to the hyoid bone.

Two tongue muscles—*hyo* and *genio-glossus*—may act as elevators of the hyoid bone: the hyo-glossus is marked with N in the Figure, and both are displayed fully in Plate xxii.

Action of the elevators. With the mouth shut and the tongue fixed against the roof, the muscles will assist in preparing the pharynx for the reception of the food, by drawing upwards and forwards the os hyoides, so as to bring the larynx under shelter of the tongue. But if the mouth is open and the tongue not in contact with the roof, the muscles are deprived of their usual point of support above, and swallowing will be performed with difficulty;—the necessary elevation of the hyoid bone in this imperfect deglutition being then dependent upon the stylo-hyoideus and posterior belly of the digastricus, which retain their usual position, and on the extreme contraction of the other muscles.

Supposing the os hyoides fixed by its depressors, the muscles used commonly as elevators of that bone will have a different action:—those attached to the jaw may then carry it downwards, so as to open the mouth; and the lingual muscles can depress the tongue.

---

* There are great variations with respect to this attachment and the state of the stylo-hyoideus muscle: some of these may be perceived in the different Plates.

ARTERIES OF THE FRONT OF THE NECK.

Only the carotid trunks and some branches of the external carotid artery are visible in front of the sterno-mastoideus.

- a. Common carotid artery.
- b. External carotid artery.
- c. Internal carotid artery.
- d. Superior thyroid artery.
- e. Lingual artery.
- f. Facial artery.
- g. Occipital artery.
- h. Posterior auricular artery.
- l. Superficial temporal artery.
- n. Internal maxillary artery.
- × Spot for ligature of the common carotid artery.

The *common carotid* trunk, a, begins opposite the sterno-clavicular articulation, and ends at the upper edge of the thyroid cartilage by splitting into two—external and internal carotid. Its situation will be marked on the surface by a line from the inner end of the clavicle to a point midway between the jaw and the ear.

Contained in a sheath of fascia with the jugular vein and the vagus nerve, it is covered throughout by muscles; and it has the following connections with the contiguous parts:—Superficial to it, besides the teguments and the platysma, are the depressors of the hyoid bone and the sterno-mastoideus—the last muscle covering it to the ending (Plate xvi.); and the others, only as high as the cricoid cartilage. Beneath the vessel is the spinal column. To its inner side lie the gullet and the air passage, with the thyroid body; and as the trachea swells out to form the larynx, necessarily the artery is carried farther from its fellow above than below. Along the outer side is a chain of lymphatic glands, which is liable to become enlarged.

The internal jugular vein, p, is parallel to, and in close contact externally with the artery; and on the left side the vein advances over it, especially lower in the neck. Three veins cross the artery;—near the chest is the anterior jugular vein, s; near the ending the upper thyroid vein; and below, ×, the middle thyroid vein.

In front of the sheath, in the upper half, the descendens noni nerve, 3, crosses obliquely from without inwards. Beneath the sheath is the sympathetic nerve; and lower down the recurrent laryngeal nerve and inferior thyroid artery cross inwards under it. In the sheath, between the artery and vein lies the vagus nerve.

No collateral offset arises commonly from the carotid artery, and the trunk remains nearly of the same size; but not unfrequently the upper thyroid branch *d* is transferred to the slight dilatation at the end.

Ligature of the vessel. The part of the common carotid marked thus × is selected for ligature because it is far removed from each end, and because it is less deep here than at a lower point. But since the vessel may bifurcate as low as the cricoid cartilage or even lower, two trunks instead of one may be met with at this spot. Should the point of splitting of the artery be recognized in the operation of ligature both trunks may be tied; but if the origin of the two trunks cannot be seen in consequence of the artery having divided very soon, the finger may be pressed on each, and that trunk may be tied which takes blood to the disease or injury for which the operation was undertaken (Quain).

Steps of the operation. With the line of the vessel in mind, the operator places the forefinger of the left hand opposite the cricoid cartilage, and makes an incision in that line three inches in length (the finger marking the centre) through the integuments, platysma, and deep fascia, down to the fibres of the sterno-mastoideus: should the cut be made too far forwards, the anterior jugular vein may be injured. Next the sterno-mastoid muscle is to be dissected back rather beyond the position of the artery, the head being inclined towards the shoulder of the same side to relax its fibres. The operator now looks for the deep guide, viz., the angle formed below by the omo-hyoid muscle, C, and the sterno-mastoideus, A, and seeks the vessel at that spot, dissecting but very little, because the descendens noni nerve, 3, and offsets of the upper thyroid vessels to the sterno-mastoideus cross the sheath.

The sheath is next to be opened towards the inner part—over the artery—avoiding the nerve, and the small vessels if possible; and after the artery has been separated from its sheath the needle is to be passed under it, whilst the opposite side of the sheath is raised with a pair of forceps. Between the artery and vein lies the vagus nerve: this is not to be included in the ligature, and if the artery has been carefully detached the nerve cannot well be caught. Before tying the thread the operator should ascertain that the pulsation in the vessel can be stopped by pressure.

Should the artery be denuded too much, the application of two ligatures may be needful—one at each end of the part laid bare. And should the size of the vein be inconveniently large, it may be diminished by the pressure of the finger of an assistant at the upper part of the wound.

In an operation on the left side the vein, from its position over the artery, would have to be turned aside; and this step may be needed also on the right side when the vein covers the artery.

In the operation on the dead body the needle will pierce readily the coats of a large loose artery, if force is used; and any check therefore to the progress of the needle in the living body should be met by drawing back the point of the instrument, and by pulling upwards tightly with a forceps the opposite side of the sheath before another attempt is made to pass the ligature.

The *internal carotid* artery, c, is the direct continuation of the common carotid trunk, and enters the skull through the temporal bone (Plate xxiv.). Below the level of the digastric muscle the artery may be reached by the surgeon, but above that muscle it is quite inaccessible. No offset is given from this vessel in the neck.

In the accessible part of its course it corresponds with the surface-line of the common carotid. It is covered, like that vessel, by the sterno-mastoideus, and rests on the spine;—at this spot it lies external or posterior to the external carotid trunk. The internal jugular vein, and the vagus and sympathetic nerves, have the same position to the internal, as to the common carotid artery.

Crossing the artery superficially is the hypoglossal nerve, 14, which sends down the descendens noni branch, 3; and beneath it the superior laryngeal nerve, 1, is directed inwards. The occipital artery, g, runs backwards over it, commonly near the digastric muscle, but sometimes lower down as in the Figure: a branch of this supplies the sterno-mastoideus.

Ligature. If this artery is ever tied it should be secured as far from its origin as it well can be; and a point between the hyoid bone and the digastric muscle may be selected as the most suitable. But the part of the artery available may be very short in consequence of the common carotid ascending as far as, or farther than the os hyoides before it splits. Should the forked ending of the common trunk be found in an operation, both the resulting arteries may be secured at their origin. Occasionally the ending of the common carotid rises still nearer to the head, and in those cases no part of the internal carotid will be below the digastric muscle.

The spot for the application of the ligature being well ascertained by means of the digastric muscle, the hyoid bone, and the line of the artery, the first steps of the operation, and the parts to be cut through are the same as those before given for ligature of the common carotid. After the sterno-mastoid muscle has been reflected the hypoglossal nerve and the occipital artery, with their branches, may be met with. When laying bare the artery care must be taken of the external carotid trunk on the one side, and of the internal jugular vein on the other; and in passing the aneurism needle the same precautions are to be observed as in the case of the common carotid (p. 147).

The *external carotid* artery, b, reaches from the upper border of the thyroid cartilage nearly to the neck of the lower jaw, and ends by dividing into temporal and maxillary branches. It is smaller than the internal carotid; and it distributes branches to the neck, and the outer parts of the head. At first it is placed on the anterior or inner side of the internal carotid, but afterwards becomes superficial to that vessel.

As high as the mastoid process the artery is covered by the sterno-mastoideus, A, the digastricus, R, and the stylo-hyoideus, P, besides the common investing superficial layers; and thence to its ending it is concealed by the parotid gland (Plate xvi.). Anterior to it are the pharynx and the jaw; and beneath it is the styloid process. No companion vein belongs to this artery.

Several nerves cross this carotid trunk:—superficial to it near the beginning is the hypoglossal nerve, 14, and near the ending the ramifications of the facial nerve, 4; whilst beneath it lie the external laryngeal, 2, the superior laryngeal, 1, and near the jaw the glosso-pharyngeal (Plate xxii.).

The offsets of the artery are numerus:—they consist of an anterior set of three, viz., thyroid, d, lingual, e, and facial, f; a posterior set of two, occipital, g, and posterior auricular, h; and an ascending set, also three in number, the temporal, l, internal maxillary, n, and ascending pharyngeal (Plate xxiv.).

Ligature. The artery is accessible below the digastric muscle, and here it is covered, like the internal carotid, by the sterno-mastoideus. Its branches are attached to the trunk so thickly as not to leave space enough between any two for the application of a ligature without the prospect of hæmorrhage when the thread comes away. Before the removal of tumors about the jaw, ligature of the external carotid trunk might be considered advisable as an auxiliary means of checking hæmorrhage during an operation.

In a wound of a branch of the external carotid the vessel should be tied, as a rule, where it is injured; but in hæmorrhage from the artery of the tongue, where the bleeding orifice cannot be secured, the surgeon may have recourse to ligature of the artery nearer the origin from the common trunk.

Branches of the carotid. The *upper thyroid, d,* ends in the thyroid body: it gives offsets to the contiguous muscles, and a laryngeal branch to the interior of the larynx with the upper laryngeal nerve, 1.

The *lingual* artery, *f*, is distributed to the tongue. Arising above or below the cornu of the hyoid bone it is directed inwards beneath the hyoglossus muscle, N. In the tongue the arteries of opposite sides converge to the tip (Plate xxii.).

If this artery is to be tied it may be secured between the origin and the edge of the hyo-glossus muscle, as it passes near the cornu of the hyoid bone. An incision directed downwards and backwards over the cornu of the os hyoides would allow the artery to be laid bare: the hypoglossal nerve is a valuable guide to the position of the vessel in the bottom of the wound.

The *facial* artery, *f*, comes off near the digastricus, and courses under it and the stylo-hyoideus, but over the submaxillary gland. It supplies branches to the surrounding parts, and a *submental* offset below the jaw.

As it crosses the jaw it lies in front of the masseter, where it is covered by the platysma: it can be easily compressed with the finger in that situation.

The *occipital* artery, *g*, begins near the digastric muscle, and is directed beneath it to the occiput: the hypoglossal nerve hooks round the vessel when this arises low down. One or more offsets enter the sterno-mastoideous.

The *posterior auricular, h,* springs near the upper border of the digastricus, and runs to the back of the ear. A cutaneous offset passes to the occiput.

Temporal and *internal maxillary.* The maxillary, *n*, courses beneath the jaw; and it will be met with in other dissections. The temporal, *l*, ascends to the side of the head, and gives offsets to the ear: anteriorly it supplies a large branch to the face—*transverse facial.*

VEINS OF THE FRONT OF THE NECK.

Three in number, the veins are named jugular—internal, external, and anterior; and they return to the chest the blood circulated through the head and neck by the carotid arteries.

p. Internal jugular vein.
r. External jugular vein.

s. Anterior jugular vein.

The *internal jugular* vein, *p*, reaches from the foramen jugulare in the base of the skull to the inner end of the clavicle, where it joins the subclavian vein (Plate xviii.). In the neck it is the companion vein to the common, and the internal carotid artery; and it is joined by the veins corresponding with the branches of the external carotid, with the exception of three which enter the external jugular.

External jugular, *r*. The course and ending of this vein have been before described (p. 131). In the Drawing the beginning of the veins by the union of the temporal and internal maxillary may be perceived: into the vein the posterior auricular branch is received lower down.

NERVES OF THE FRONT OF THE NECK.

Several cranial nerves appear in the region dissected; and they are distributed to the face, the tongue, the windpipe, and the gullet. Only one spinal nerve is seen.

1. Upper laryngeal nerve.
2. External laryngeal nerve.
3. Descendens noni nerve.
4. Facial nerve.
5. **Temporo-facial piece of the facial nerve.**
6. **Cervico-facial piece of the facial nerve.**
7. **Nerve to the digastric and stylo-hyoid muscles.**
8. Posterior auricular nerve.
9. Branches of the great auricular nerve joining the facial.
10. Great auricular nerve.
11. Auriculo-temporal branch of the fifth nerve.
12. Mylo-hyoid branch of the fifth nerve.
13. Spinal accessory nerve.
14. Hypoglossal nerve.

The *facial* nerve, 4, issuing from the skull by the stylo-mastoid foramen, divides into two chief parts,—temporo-facial, 5, and cervico-facial, 6: these pass forwards through the parotid gland to the forehead, the face, and the superficial parts of the neck as low as the hyoid bone.

As soon as the nerve leaves its bony canal it gives off the posterior auricular nerve, 8, and a muscular branch, 7, to the posterior belly of the digastricus and to the stylo-hyoideous.

It is chiefly a motor nerve; and it gives the ability to contract to the superficial muscles of the head, ear, face, and neck. It joins freely with the sensory fifth nerve, and furnishes offsets also to the integuments; and as it supplies alone the posterior belly of the digastricus and the stylo-hyoideus, it must confer on them sensibility as well as motion.

The *hypoglossal* nerve, 14, is the motor nerve of the tongue. Descending through the neck with the great bloodvessels till it comes below the digastric muscle, it is then directed forwards over the carotids to the submaxillary region: it will be continued in Plate xxii.

As it crosses the vessels it supplies two offsets:—one, *descendens noni*, 3, enters the depressor muscle of the hyoid bone, after joining with the spinal nerves (Plate xxiv.); the other, much smaller, ends in the thyro-hyoideus, F.

Branches of the vagus. Two branches of this nerve, viz., the upper laryngeal, 1, and the external laryngeal, 2 (an offset of the first), are furnished to the larynx; their distribution will be referred to more fully in the description of the Plate of the larynx.

Branches of the fifth nerve. The *auriculo-temporal*, 11, is a sensory nerve, and ascends with the temporal artery to the side and top of the head; it supplies the ear with the attrahent muscle, and the parotid gland. The *mylo-hyoid* branch, 12, lies below the jaw, and ends in the anterior belly of the digastricus, and the mylo-hyoideus: contractility and sensibility are given to those muscles by the nerve.

The *great auricular* nerve, 10, of the cervical plexus is displayed in Plate xvi. In this Figure the communicating branches, 9, through the parotid to the facial nerve are brought into view.

PLATE X

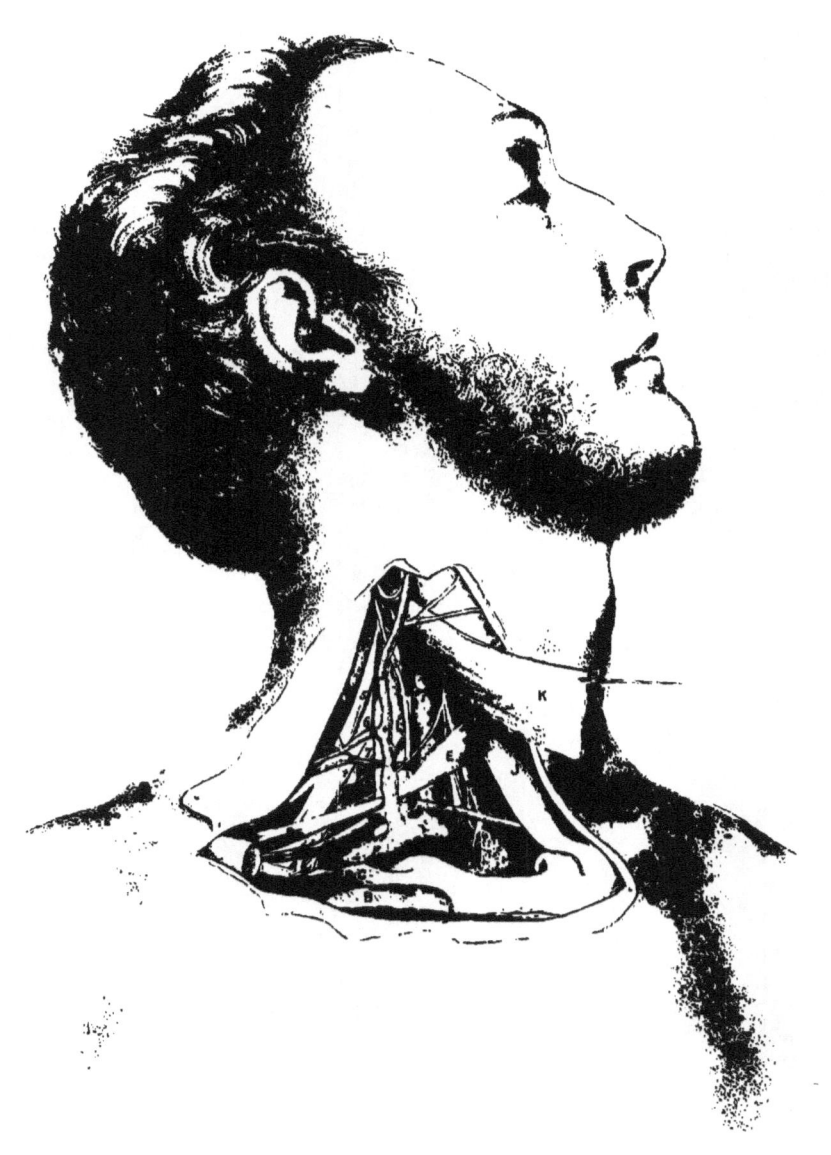

DESCRIPTION OF PLATE XVIII.

THE dissection of the subclavian bloodvessels with the contiguous nerves and muscles is portrayed in this Plate.

This view has been obtained by cutting through the sterno-mastoideus muscle, after the dissection of the posterior triangular space; and by sawing through the clavicle and removing the inner end. On the section of the clavicle the shoulder falls back, and the subclavius and omo-hyoideus muscles are stretched.

MUSCLES OF THE SUBCLAVIAN REGION.

Only the subclavius, the posterior belly of the omo-hyoideus, and the anterior scalenus will be now referred to, the other muscles having been described in other dissections.

- A. Pectoralis major, cut.
- B. Intercostal muscles of the first space.
- C. Subclavius muscle.
- D. Omo-hyoideus—posterior belly.
- E. Omo-hyoideus—anterior belly.
- G. Sterno-hyoideus.
- H. Sterno-thyroideus.
- J. Sternal part of the sterno-mastoideus.
- K. Clavicular part of sterno-mastoideus, cut.
- L. Anterior scalenus.
- N. Middle scalenus.

Anterior scalenus, L. The connections of the muscle may be here studied: the attachments are given at p. 123. It lies beneath the sterno-mastoid and omo-hyoid muscles; and it is connected with the following vessels and nerves. In front of it lies the subclavian vein, *p*, with the external jugular, *s*, and anterior jugular, *v;* and along the inner edge descends the large internal jugular vein, *r*. Beneath it is the subclavian artery, *b*, and on it are three small arteries, supra-scapular, *l*, transverse cervical, *h*, and ascending cervical, *f*. Issuing from beneath the muscle are the large cervical nerves; and running down in front of it is the phrenic nerve, 3.

Omo-hyoideus, B. The posterior belly of this muscle is attached behind to the upper border of the scapula, and ends in front in a tendon beneath the sterno-mastoideus. It receives a small vessel from the supra-scapular, and a nerve from the descendens noni; and the supra-scapular vessels, *l* and *w*, and the supra-scapular nerve, 9, course backwards with it. See also Plate xv., and p. 123.

Subclavius muscle, C. In Plate ii. this may be viewed in its natural state, surrounded by a sheath of fascia. It arises from the first rib where the bone and cartilage join; and it is inserted into the grooved under surface of the clavicle. The inner part of the muscle shows a ragged edge, where was detached from the bone.

THE SUBCLAVIAN ARTERY AND ITS BRANCHES.

The subclavian artery runs through the lower part of the neck, and gives branches to the chest, the shoulder, the neck, and the brain.

a. First part of the subclavian trunk.
b. Third part of the subclavian.
c. Common carotid artery.
d. Inferior thyroid artery.
f. Ascending cervical artery.
h. Transverse cervical artery.
l. Supra-scapular artery.
n. Internal mammary artery.

The *subclavian* artery of the right side begins opposite the inner end of the clavicle, where the innominate trunk bifurcates, and ends at the lower border of the first rib by becoming axillary. Between those points the artery forms an arch with the convexity upwards, which lies between the scaleni muscles. Its numerous connections will be best learnt by dividing the trunk of the artery into three parts:—one inside, one beneath, and one outside the anterior scalenus.

The *first part* of the artery, *a*, is concealed by the muscles of the front of the neck, viz., sterno-mastoideus, J, sterno-hyoideus, G, and sterno-thyroideus, H; also by the integuments and the platysma. It lies deeply, but not in contact with the spinal column.

Lying near the chest and below the artery is the arch of the subclavian and innominate veins; and crossing it at right angles is the internal jugular vein, *r*, with the vertebral vein beneath this. And in front of the artery though separated by muscles is the anterior jugular vein, *v*.

The vagus nerve, 10, crosses over the artery inside the jugular vein, together with some branches of the sympathetic; and the recurrent branch of the vagus, and the cord of the sympathetic, lie beneath it (Plate xxiv.).

The *second* or *middle part* of the artery, the shortest and highest, is covered by the anterior scalenus, L, and the sterno-mastoideus, K; and rests on the middle scalenus, N.

No vein touches the artery in the second part, for the anterior scalenus intervenes between the subclavian vein and artery. Two arteries, transverse cervical, *h*, and supra-scapular, *l*, lie near the line of the subclavian trunk, the former being rather above, and the latter below it.

The lower cervical nerves are above the vessel between the scaleni; and the phrenic, 3, crosses it, but separated by the scalenus anticus.

The *outer* or *third part*, *b*, is the most superficial, and decends over the first rib to the axillary space, crossing beneath the omo-hyoideus, D, the subclavius, C, and the clavicle. This part appears in the posterior triangular space of the neck (Plate xv.); and its connections are described in p. 126.

Into the concavity of the arch of the bloodvessel the bag of the pleura projects, for this membrane rises above the first rib, and comes in contact with the first and second parts of the subclavian artery: this connection of the serous membrane must be remembered when ligature of the second part of the artery is to be undertaken. Alterations affecting the arch have been dwelt on in p. 129.

Number and *position of the branches*. Usually four branches arise from the artery in the following manner;—three are connected with the first part, and one with the second part; whilst no branch, as a rule, comes from the third part. Very commonly, however (Quain), an offset (posterior scapular) of the branch. *h*, is attached to the last part of the subclavian trunk.

From the position of the branches, the connections, and the difference in the depth of the ends of the subclavian trunk, the third or external part, *b*, will be best suited for ligature on account of its comparative freedom from any branch, and its easily accessible position. As the second part gives origin commonly to but one branch it may admit of being tied under some circumstances. Whilst the inner or first part is so beset by branches as not to possess commonly an interval sufficient for the application of a ligature without secondary hemorrhage. On the left side the complicated connections forbid the attempt to put a thread on the first part.

Ligature. The steps of the operation for securing the artery in the third part, or beyond the scalenus, have been detailed at p. 129.

Should the less usual operation of tying the second part of the artery be resorted to, the clavicular piece of the sterno-mastoideus and the anterior scalenus would have to be cut through. In dividing the scalenus great care should be taken of the phrenic nerve, 3, on its front. Ordinarily the external jugular vein lies outside the scalenus: with the position here taken it would need to be cut through, and the ends would require to be tied.

Branches of the subclavian artery. At their origin the branches are concealed by the jugular vein and the anterior scalenus, but in Plate xxiv. most may be seen. From the first part come the vertebral, the thyroid axis, and the internal mammary; and from the second part, the upper intercostal, with a small branch to the spinal canal (Quain).

1. The *vertebral* is the first branch, and ascends to the brain through the apertures in the six upper cervical vertebræ.

2. The *thyroid axis*, a short thick trunk, splits into the three following;—*Inferior thyroid, d.* This is a tortuous artery, and ends in the thyroid body: an offset, the *ascending cervical, f,* lies between the anterior scalenus and the larger anterior rectus, supplying offsets to both, and to the spinal canal. The *transverse cervical, h,* crosses the scalenus, and ends under the trapezius by dividing into two. The *supra-scapular, l,* courses along the clavicle to the scapula, on the dorsum of which it ramifies.

3. The *internal mammary, n,* arises opposite the vertebral and beneath the jugular vein: it enters the thorax through the upper opening, and is continued to the wall of the abdomen.

The *superior intercostal* (intercosto-cervical) arises under the scalenus: it supplies offsets to the upper two intercostal spaces; and a large branch to the back of the neck (deep cervical), which is delineated in Plate xix.

SUBCLAVIAN AND JUGULAR VEINS.

The veins of the arm and of the same side of the neck meet at the top of the thorax, and blend in one large trunk—the innominate: the limb vein is called subclavian, and the neck veins jugular.

p. Subclavian vein.
r. Internal jugular vein.
s. External jugular vein.

t. Transverse cervical vein.
v. Anterior jugular vein.
w. Supra-scapular vein.

The *subclavian* vein, *p*, is rather shorter than its corresponding artery, and ends near the inner border of the scalenus by joining the internal jugular to form the innominate trunk. Arched like the artery, it is placed in front of the scalenus, and commonly below the level of the clavicle. Valves exist in the trunk outside the place of entrance of the external jugular, *s*.

The veins joining it are the external and anterior jugular, and the vertebral. At the back of the vein, near the internal jugular, the right lymphatic duct opens; and at a similar spot on the left side the thoracic duct is received.

External and *anterior jugular veins.* The ending of these veins is seen in this Plate, and their course is described in p. 131. The external jugular, *r*, receives the transverse cervical branch, *t*, and the supra-scapular *w*, and joins the subclavian vein outside the scalenus anticus. The anterior jugular, *v*, enters either the subclavian vein or the external jugular: when this vein is tributary to the external jugular it wants valves (Struthers).

Internal jugular vein, r. The lower dilatation of the vein is laid bare. Before its junction with the subclavian it is narrowed, and at the less wide part is a pair of valves to prevent the blood rushing backwards to the neck.*

The *innominate* is the large venous trunk formed by the union of the subclavian and internal jugular veins: it enters the chest, and joins with its fellow in the upper cava. The connections of the vein in the neck may be studied in the Figure.

NERVES OF THE SUBCLAVIAN REGION.

Most of the nerves are continued to distant parts, only two being distributed to the neighboring muscles.

---

* These valves were first described by Dr. Struthers. See an account of them in the Edinb. Méd. Journal for Nov., 1856, p. 241.

1. Great auricular nerve.
2. Superficial cervical nerve.
3. Phrenic nerve.
4. Descendens noni nerve.
5. Fifth cervical nerve.
6. Sixth cervical nerve.
7. Seventh cervical nerve.
8. Eighth cervical nerve.
9. Supra-scapular nerve.
10. Vagus nerve.
† Nerve to the subclavius.

The *diaphragmatic* (phrenic) nerve, 3, springs from the fourth spinal nerve in the cervical plexus, and is sometimes connected with the fifth spinal as it passes by that trunk. In the neck it courses over the anterior scalenus muscle, crossing from the outer to the inner edge; and entering the chest beneath the innominate vein, it is transmitted through that cavity to the diaphragm. It is the motor nerve of the diaphragm.

Descendens noni nerve, 4. For the beginning of this branch of the hypoglossal, see Plate xvii. At the lower part of the neck it ends in branches for the sterno-hyoideus, G, sterno-thyroideus, H, and the posterior belly of the omo-hyoideus, D, as well as the anterior belly of the same muscle.

Brachial plexus. The lower four cervical nerves, 5, 6, 7, 8, join with the first dorsal to form the plexus. The branches of the plexus above the clavicle are enumerated in p. 134; but only two, nerve to the subclavius, †, and the supra-scapular, 9, are seen in a front-view of this region.

The *vagus nerve*, 10, passes through the neck and thorax to the belly. At the lower part of the neck, on the right side, it occupies the interval between the jugular vein and the carotid artery, and crosses over the subclavian artery but beneath the innominate vein.

It furnishes a small cardiac branch near the subclavian artery; and close below that vessel it sends backwards the recurrent or inferior laryngeal nerve.

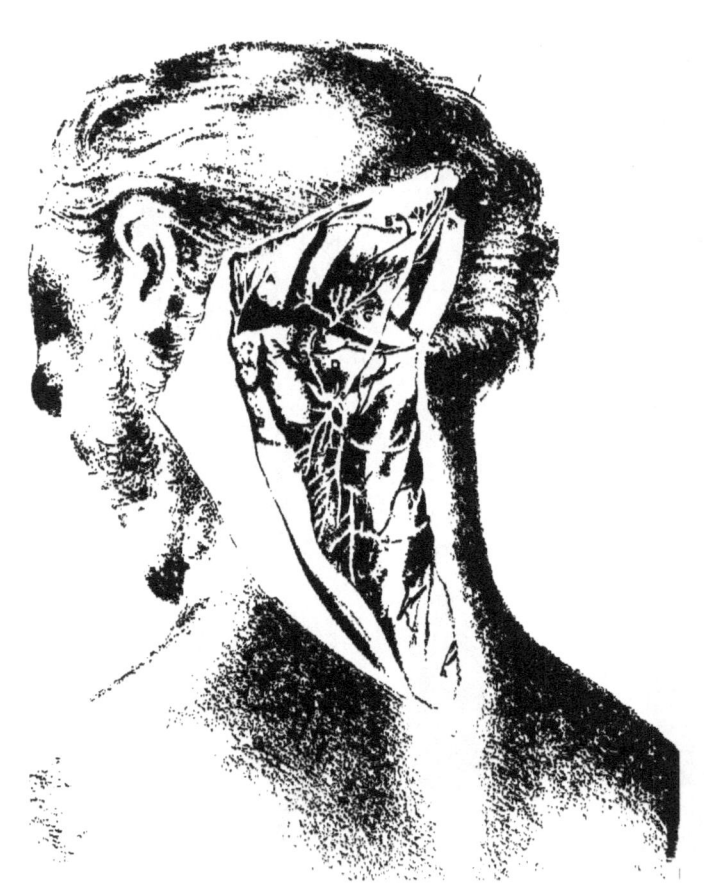

DESCRIPTION OF PLATE XIX.

A VIEW of the deep muscles, and of the vessels and nerves at the back of the neck, is here given.

After the integuments and the superficial muscles have been reflected the complexus is to be divided near the head; and this last muscle being thrown down and out, the vessels and nerves are to be sought in the dense tissue and fascia in which they are imbedded. Lastly, the muscles are to be defined.

DEEP MUSCLES OF THE NECK.

Extensor and rotator muscles of the head and neck lie beneath the complexus, B. Between the head and the first two vertebræ, and corresponding with the interspinales, are placed the recti muscles; and laterally there are two other small muscles, the obliqui. Occupying the vertebral groove is the semispinalis colli.

A. Sterno-mastoideus.
B. Complexus cut through.
C. Semispinalis colli.
D. Obliquus inferior.

F. Obliquus superior.
G. Rectus posticus major.
H. Rectus posticus minor.

The *complexus muscle*, B, is attached by the outer edge to the transverse processes of the upper dorsal vertebræ, and to the articular processes of the cervical vertebræ, except the first two; and by the inner edge it is connected with the spines of one or two lower cervical and upper dorsal vertebræ. It is inserted into the mid part of the occipital bone between the curved lines.

Towards the inner edge a piece of the muscle possesses a middle tendon, and this is often described separately as the *biventer cervicis*.

If the muscles of both sides act they will maintain the head erect, or will bring it back (raising the face) according to the degree of contraction; but supposing only one to contract, the occiput will be inclined down and out towards the transverse processes of the same side.

Semispinalis colli, C. Filling the vertebral groove with the multifidus spinæ, it is attached externally, like the preceding muscle, to the transverse processes of the upper dorsal vertebræ, and to the articular processes of the cervical vertebræ, except the first three; and internally it is inserted into the spines of the cervical vertebræ below the first.

Acting with its fellow it extends the spine: by itself, it rotates the spine, turning the face to the opposite side.

The *obliquus inferior*, D, slants between the first two vertebræ: it arises from the spine of the axis, and is inserted into the transverse process of the atlas.

Drawing backwards the lateral part of the atlas it rolls this bone round the odontoid process of the axis, and rotates indirectly the head, moving the face to its own side.

The *obliquus superior*, F, arises from the transverse process of the atlas, where the preceding is attached, and is inserted into the occipital bone between the curved lines, and near the mastoid process.

The muscle can draw back the head; and may check a too great forward movement, as in nodding.

The *rectus posticus major*, G, arises from the spine of the second vertebra; and widening as it ascends obliquely, it is inserted into the outer half of the lower curved line of the occipital bone, where it is partly concealed by the obliquus superior.

This muscle extends the head, and brings the face to its own side by moving the atlas round the odontoid process of the axis.

Rectus posticus minor, H, arises from the arch of the atlas, close to the middle line; and is inserted into the inner half of the lower curved line of the occipital bone. The muscle extends the head.

ARTERIES OF THE BACK OF THE NECK.

Three arteries supply the back of the neck, and connect the vessels of the head with those of the trunk. In the neighborhood of the thorax small offsets of the dorsal arteries appear.

a. Occipital artery.
b. Deep cervical branch of the occipital.
c. Offset to the small rectus muscle.
d. Vertebral artery.
e. Cervical branch of the vertebral.
f. Anastomosis of the vertebral and deep cervical arteries.
g. Deep cervical artery.
h. Dorsal arteries—the inner branches.

The *occipital artery, a*, courses to the integuments of the back of the head over the obliquus superior and the complexus, and beneath the sterno-mastoideus, the splenius, and the trachelo-mastoideus: near the middle line it pierces the trapezius.

It furnishes a *cervical* branch, *b*, to the neck (ram. princeps cervicis), which descends beneath the complexus, B, supplying the deep muscles, and anastomoses with branches of the vertebral and deep cervical arteries. An offset passes over the complexus, and supplies the superficial muscles.

The *vertebral artery, d*, in its course to the interior of the skull is directed backwards in a groove on the neural arch of the atlas. Lying deeply in the bottom of the hollow between the large rectus and the oblique muscles, it furnishes one or two muscular offsets, *e*, and communicates with the contiguous arteries.

The *deep cervical artery, g*, is the dorsal offset of the upper intercostal (p. 156), and reaches the back of the neck by passing between the transverse processes of the last cervical and first dorsal vertebræ. At the back of the neck it ascends under the complexus as high as the axis, where it communicates with the two arteries before described. It supplies chiefly the complexus and the semispinalis colli.

The height at which the artery appears is very uncertain; and it may be represented by two branches of different arteries. In obstruction of the circulation in the common carotid the blood will be conveyed to the exterior of the head by means of the anastomosis between the profunda and the occipital artery.

The *companion veins* of the arteries have not been included in the Plate: they resemble the arteries, with the exception of the vertebral which begins on the back of the head and neck, and does not enter the skull.

NERVES OF THE BACK OF THE NECK.

The anatomy of the posterior primary branches of the cervical nerves beneath the complexus is here shown. A part of the small occipital nerve appears behind the ear.

1. First or suboccipital nerve.
2. Second cervical nerve.
3. Third cervical nerve.
4. Fourth cervical nerve.
5. Fifth cervical nerve.
6. Sixth cervical nerve.
7. Seventh cervical nerve.
8. Small occipital nerve.

The *first nerve*, 1, appears beneath the vertebral artery, and ends in branches to the complexus, and the recti and obliqui muscles: it is joined to the second nerve by a loop.

Other cervical nerves. The remaining seven cervical nerves divide into two—inner and outer branches, as soon as they leave the spinal canal.

The *external branches* are not laid bare except that of the second: they are small, and are distributed to the muscles outside the complexus, viz., splenius, cervicalis ascendens, and transversalis colli and trachelo-mastoideus.

The *internal branches* are directed inwards—the four highest over, and the remaining three through the semispinalis colli; and at the spines of the vertebræ those that lie on the semispinalis become cutaneous. They supply the complexus and the muscles filling the vertebral groove, with the interspinales. The following are the chief differences in these nerves:—

The branch of the second nerve, 2, the largest of all, pierces the complexus and trapezius, and becoming cutaneous is distributed to the occiput: it is named *great occipital*, and is joined by the small occipital nerve, 8. It supplies branches to the inferior oblique and complexus muscles; and it communicates by loops with the first and third nerves.

The cutaneous part of the third nerve, larger than those below it, sends upwards a branch to the occiput, which joins the larger occipital nerve.

The connecting pieces between the inner branches of the first three nerves are sometimes absent. M. Cruveilhier describes this looped arrangement as the posterior cervical plexus.

The *small occipital nerve*, 8, is an offset of the cervical plexus (Plate xv.). it ends in the integuments of the occiput, and joins the great occipital nerve.

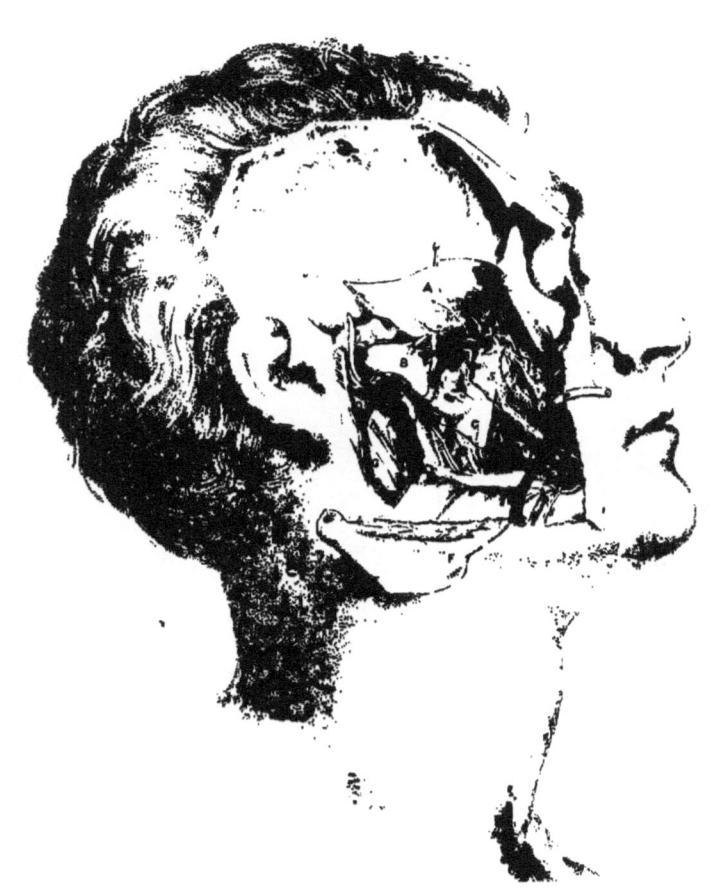

DESCRIPTION OF PLATE XX.

In this dissection of the pterygoid region the muscles of mastication, and the internal maxillary artery with its branches can be studied. Most of the branches of the inferior maxillary nerve come also into sight.

This superficial dissection will be made by detaching and throwing down the zygomatic arch with the masseter muscle, by sawing off and raising the coronoid process with the temporal muscle: and by removing the piece of the ramus of the jaw between the condyle and dental foramen. After each sawing of the bone the fat is to be carefully removed.

MUSCLES OF MASTICATION.

The muscles employed in mastication are attached chiefly to the angle and processes at the back of the lower jaw; but one, which occupies the cheek, blends with the lip-muscles.

A. Temporal muscle.
B. External pterygoid muscle.
C. Internal pterygoid muscle.
D. Buccinator muscle.
F. Masseter muscle.
G. Digastric muscle (posterior belly).

H. Stylo-hyoideus.
L. Stylo-glossus.
N. Internal lateral ligament.
O. Styloid process.
P. Duct of the parotid gland.

The *temporal muscle*, A, arises from the temporal fossa on the side of the skull, and from the upper part of the temporal fascia; and the fibres converge to a tendon which is attached to the under surface of the coronoid process, and to the groove along the fore part of the ramus of the jaw.

Comparatively superficial above, the muscle passes below beneath the zygomatic arch and the masseter muscle, and rests on the external pterygoid, B. Near the zygoma a stratum of fat intervenes between the fleshy fibres and the temporal fascia.

In mastication this muscle crushes the food by raising the lower jaw; and if the jaw has been moved forwards, the hinder fibres may be able to bring that bone backwards, with the aid of the muscles attached to the chin.

The *masseter muscle*, F, is placed external to the ramus of the jaw. It takes origin from the lower border and inner surface of the zygomatic arch; and it is inserted into the outer surface of the ramus of the jaw, from the tip of the coronoid process to the angle, and as far forwards as the second molar tooth. The superficial fibres take a direction down and back across the deeper and straighter fibres.

This muscle is the external elevator of the angle of the jaw.

The *internal pterygoid muscle*, C, has a position inside the ramus of the jaw similar to that of the masseter outside. The muscle arises chiefly from the pterygoid fossa, but below from the palate and upper jaw bones by a process which extends in front of the lower part of the external pterygoid muscle. It is inserted into the inner surface of the angle and ramus of the jaw as high as the dental foramen.

It raises the angle of the jaw in conjunction with the masseter, and may be considered the internal elevator of the angle.

The *external pterygoid* muscle, B, is directed almost horizontally back and out from the base of the skull to the condyle of the jaw. Arising from the outer surface of the external pterygoid plate, and from the contiguous part of the great wing of the sphenoid bone as high as the crest, the muscle is inserted into the front of the neck of the lower jaw, and into the interarticular fibro-cartilage.

An interval separates the attachments to the external pterygoid plate and the great wing, through which the internal maxillary artery, d, usually passes.

If the muscles of both sides act the jaw is moved downwards and forwards, and the front lower teeth pass beyond the upper. If only one muscle acts, say the right, it draws the condyle of the same side further into the articular socket, and causes the chin to project to the left of the middle line of the head, the grinding teeth of the lower jaw passing laterally over those of the upper jaw.

The *buccinator* muscle, D, forms a thin fleshy layer in the cheek between the mucous membrane and the teguments. It is attached to the jaws opposite the molar teeth, and between the jaws at the back of the mouth to a fibrous band—the pterygo-maxillary ligament. Towards the corner of the mouth the fibres are aggregated together, and entering the lips blend with the orbicularis oris muscle.

In the movements of the lips the muscle retracts the corner of the mouth, and so widens that aperture, and wrinkles the cheek.

In mastication it is applied to the jaws, and prevents the food escaping outside the teeth; when it is paralyzed the food distends it and the cheek in an inconvenient manner.

In playing a wind instrument this muscle is flattened, and the fibres are contracted for the purpose of driving the outgoing air through the channel of the mouth; but in the use of a blow-pipe the muscle is distended because the mouth is used as a reservoir, but the fibres contract at the same time, to maintain a continuous and active current of air.

INTERNAL MAXILLARY ARTERY.

The chief vessel in this dissection is the internal maxillary artery, which is continued through the pterygoid region to the deep parts of the head, the nose, and the palate, supplying many offsets.

- a. External carotid artery.
- b. Posterior auricular branch.
- c. Superficial temporal artery.
- d. Internal maxillary artery.
- e. Inferior dental branch.
- f. Branch with the gustatory nerve.
- g. Deep temporal artery.
- h. Buccal artery.
- l. Posterior dental branch.
- n. Facial artery.
- r. Inferior labial branch.
- s. Masseteric branch, cut.

The *internal maxillary* artery, d, is one of the terminal branches of the external carotid, and runs upwards and inwards over or under the external pterygoid muscle to the spheno-maxillary fossa, where it ends in branches for the nose, the palate, and the pharynx. It gives numerous branches, and these are classed into three sets:—one external to the pterygoid muscle, another whilst the artery lies on the muscle, and a third internal to the muscle, or in the spheno-maxillary fossa. The first two sets will be mainly referred to now.

The *first set* of *branches*, two in number (dental and meningeal), belong to the lower jaw and the skull.

The *inferior dental* artery, e, enters the canal in the lower jaw with the nerve of the same name, and supplies the teeth and the lower part of the face. Before it enters the bone, a small offset (mylo-hyoid) descends with a fine nerve in a groove inside the ramus of the jaw.

The *large* or *middle meningeal* artery arises opposite the preceding, and is concealed by the external pterygoid: it is delineated in Plate xxi., *b*.

A third small artery, *f*, which has not been described by Anatomists, runs with the gustatory nerve, and supplies the cheek, and the floor of the mouth external to the tongue.

The *second set* of *branches* is distributed to the muscles of mastication as below:—

The *deep temporal*, *g*, two in number, enter the fore and hinder parts of their muscle. The *masseteric* branch, *s*, springs in common with the posterior temporal, and enters the hinder border of the masseter: it has been cut in the removal of the muscle. The *buccal* branch, *h*, descends to the cheek and the buccinator muscle: it anastomoses with the facial artery. Branches to the pterygoid muscle are shown in Plate xxi.

Third set of branches. Only one of these branches, the *posterior dental*, *h*, is seen in the dissection. It takes a tortuous course to the front of the upper jaw, where it communicates with the infra-orbital: it will be given more fully in Plate xxiii.

The *facial artery*, *n*, also a branch of the external carotid (Plate xvii.), is displayed as it crosses the jaw. It ascends with a wavy course to the root of the nose, passing near the corner of the mouth.

Named branches supply the lips and the nose, and one of these to the lower part of the face is the *inferior labial*, *r*. Unnamed branches ramify in the cheek, and anastomose with the buccal and transverse facial arteries.

MAXILLARY AND FACIAL VEINS.

t. External jugular vein.
v. Superficial temporal.
w. Internal maxillary vein.

x. Facial vein.
z. Deep facial, or anterior internal maxillary.

The *facial vein*, *x*, begins near where the companion artery ceases, and crosses the face to the jaw; but it takes almost a straight line from the root of the nose to the front of the masseter muscle, and does not follow the windings of the facial artery. It ends in the neck in the internal jugular trunk.

Besides branches received from the orbit and the face, it is joined opposite the angle of the mouth by a vein—the *deep facial, z*, or the anterior internal maxillary, which brings blood from the pterygoid region and the upper jaw.

Internal maxillary vein, w. Only the ending of this in the external jugular remains,—the plexiform continuation of it by the side of the artery having been taken away.

NERVES OF THE PTERYGOID REGION.

The nerves appearing in this dissection are branches of the inferior maxillary trunk of the fifth cranial nerve, with the exception of two small nerves, one lying along the upper jaw, and another on the lower jaw.

1. Auriculo-temporal nerve.
2. Inferior dental nerve.
3. Gustatory nerve.
4. Masseteric nerve, cut.
5. Buccal nerve.
6. Posterior dental nerve.
8. Buccal branches of the facial nerve.

The anatomy of the inferior maxillary nerve is described with Plate xxi.; but the position of its several branches passing the external pterygoid can be here seen before the muscle is raised.

This large trunk of the fifth nerve is concealed as it leaves the skull by the external pterygoid; and its branches escape through the muscle or at its edges. Appearing at the upper border are the masseteric nerve, 4, and the deep temporal (Plate xxi., 8); and issuing at the lower border are three large trunks, viz., the auriculo-temporal, 1, the dental, 2, and the gustatory, 3. The *buccal nerve,* 5, comes forwards between the two pieces of the pterygoideus externus.

The *posterior dental nerve,* 6, a branch of the upper maxillary trunk, descends along the upper jaw with its artery: its origin and distribution may be referred to in Plate xxiii.

DESCRIPTION OF PLATE XXI.

This Illustration of the deep dissection of the pterygoid region exhibits the third trunk of the fifth cranial nerve, and the deep branches of the internal maxillary artery.

In preparing the dissection the internal maxillary artery should be cut through, and the condyle of the jaw having been disarticulated should be drawn forwards with the external pterygoid muscle. After the removal of the fat the nerves and vessels will be ready for learning.

MUSCLES OF MASTICATION.

The muscles described with Plate xx. are met with again in this view, and they are marked with the same letters of reference. A better idea of the wide origin of the external pterygoid is obtained in this Plate.

- A. Temporal muscle.
- B. External pterygoid muscle.
- C. Internal pterygoid muscle.
- D. Buccinator muscle.
- F. Masseter muscle.
- G. Digastric muscle.
- H. Zygoma thrown down.
- L. Condyle of the jaw.
- N. Internal lateral ligament.
- O. Styloid process, and stylo-maxillary ligament.

INTERNAL MAXILLARY ARTERY.

The meningeal and the muscular branches of the internal maxillary artery, which were hidden in Plate xx., are now brought under notice; and the other arteries, which are the same as in the preceding Figure, are marked by the same letters.

- *a.* External carotid trunk.
- *b.* Large meningeal artery.
- *c.* Small meningeal branch.
- *d.* Internal maxillary artery.
- *e.* Inferior dental branch.
- *f.* Branch with the gustatory nerve.
- *g.* Deep temporal branches.
- *h.* Buccal branch.
- *l.* Posterior dental branch.
- *n.* Facial artery.
- *t.* External jugular vein.

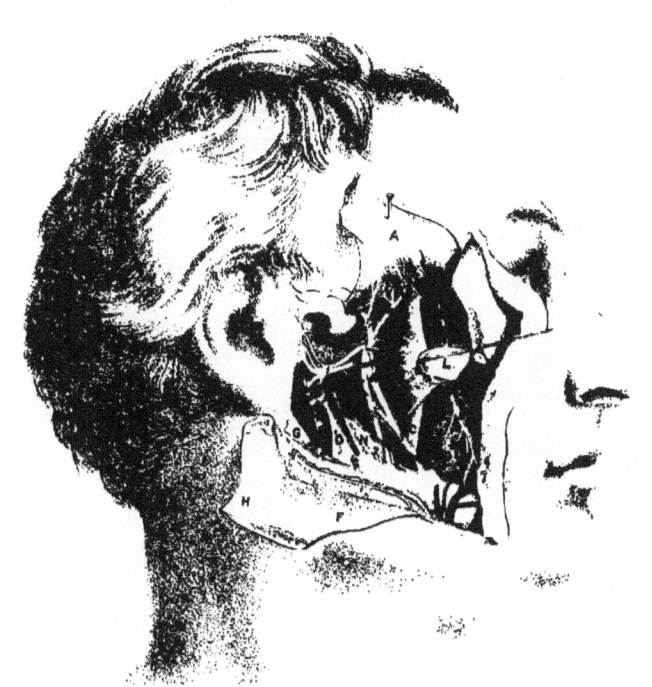

The *large* or *middle meningeal* artery, *b*, ascends to the head beneath the external pterygoid muscle, and enters the skull through the foramen spinosum (p. 110). It supplies branches to the temporal and external pterygoid muscles, an offset to the tympanum through the Glaserian fissure, and the following:—

Small meningeal branch, *c*. Arising from the large meningeal, it enters the skull through the foramen ovale: an offset is furnished outside the skull to the internal pterygoid with the branch of nerve to that muscle.

INFERIOR MAXILLARY NERVE.

The branches of the inferior maxillary nerve, whose lettering corresponds with that in Plate xx., are here traced backwards to the foramen of exit of their trunk from the skull.

1. Auriculo-temporal nerve.
2. Inferior dental nerve.
3. Gustatory nerve.
4. Masseteric branch, cut.
5. Buccal branch.
6. Chorda tympani nerve.
7. Mylo-hyoid branch.
8. Deep temporal branch.
9. Branch to the external pterygoid.
† Branch to the internal pterygoid.

The *inferior maxillary* or the third trunk of the fifth cranial nerve (Plate xiii.) leaves the skull by the foramen ovale, and splits at once into two under the external pterygoid muscle, viz.—an anterior small part, and a posterior large part. And as the nerve is composed of a motor and a sensory root (p. 108), the function bestowed by its offsets will be determined by their receiving filaments from only one or from both roots.

The *small piece* of the nerve breaks up into branches to most of the muscles of mastication as below:—

The *masseteric branch*, 3, courses above the pterygoideus externus and through the sigmoid notch to the under surface of its muscle, in whose fibres it can be followed nearly to the anterior edge: it gives an offset to the back of the temporal muscle.

The *deep temporal* branch, 8, is directed upwards on the skull into the fibres of the temporal muscle, and usually with an artery of the same name.

The *buccal branch*, 5, pierces the external pterygoid, and is continued over the buccinator towards the corner of the mouth; it supplies chiefly the buccinator muscle as well as the integuments covering, and the mucous membrane lining the same. In the cheek it joins in a plexus, *buccal*, with the facial nerve (Plate xx., 8). Two masticatory muscles, viz., the external pterygoid and the temporal, receive offsets from this branch.

A *branch* to the *pterygoideus externus*, 9, enters the under surface of that muscle.

This smaller part of the inferior maxillary nerve contains portions of both roots; these are disposed in a peculiar way, and give different functions to the branches. Thus the nerves furnished by it to the jaw muscles—masseter, temporal, and external pterygoid—are constructed from both roots, like spinal nerves, and give sensibility and contractility to those muscles. The nerve to the buccinator on the contrary is formed altogether by the sensory root, and bestows only sensibility on the muscle and the other parts to which it is distributed.

The *larger piece* of the *inferior maxillary* nerve ends in three good-sized trunks, and gives a branch to the internal pterygoid muscle.

The *auriculo-temporal* nerve, 1, beginning generally by two roots, is inclined backwards beneath the external pterygoid muscle, and ascends finally with the temporal artery to the integuments of the side of the head. It communicates largely with the facial nerve; and it supplies also the articulation of the jaw, the meatus of the ear, and the parotid gland.

The *inferior dental* nerve, 2, descends over the pterygoideus internus and the internal lateral ligament to the dental foramen of the lower jaw, and is distributed to the teeth, and the lower part of the face.

A small muscular branch, *mylo-hyoid*, arises from the nerve near the jaw, and runs in a groove in the bone to the anterior belly of the digastricus, and the mylo-hyoideus (Plate xvii.).

The *gustatory nerve*, 3, is directed downwards to the front of the internal pterygoid muscle, near the attachment to the jaw: its further course in the tongue will be represented in Plate xxii. Under the external pterygoid muscle it is joined by the chorda tympani nerve, 6.

The *branch* to the *internal pterygoid* muscle, †, comes from the large part of the inferior maxillary trunk, and enters the under surface of its muscle. Around the root of this branch, and on the inner or deep surface of the large trunk, lies the otic ganglion, which furnishes offsets to two other muscles, viz., the tensor tympani and the circumflexus palati: this body can be recognized only in a view from the inner side.

The large part of the inferior maxillary trunk receives fibrils from both roots of the fifth nerve, like the smaller piece; but as the part contributed by the sensory root is much the largest, most of the branches are formed by this alone, and are therefore sensory in function. The three large trunks, auriculo-temporal, 1, dental, 2, and gustatory, 3, are solely sensory nerves; and the last is one of the nerves of taste. The muscular branches receiving offsets from both roots, bestow sensibility and contractility on the muscles before mentioned, viz., the pterygoideus internus, the mylo-hyoideus, the anterior belly of the digastricus, the circumflexus palati, and the tensor tympani.

The *chorda tympani nerve*, 6, is a branch of a motor nerve—the facial (p. 108), and issues from the cranium through, or by the side of the Glaserian fissure. It is applied to the gustatory under the external pterygoid muscle, and is conveyed by that nerve trunk to the tongue, where it is distributed: at the point of contact one or two offsets join the gustatory.

The two following pieces of fascia, which are called ligaments, look like distinct bands in consequence of the removal of the rest of the cervical fascia, with which they are continuous.

The *internal lateral ligament* of the articulation of the jaw N, is attached by one end to the base of the skull, and by the other to the margin of the dental foramen, and to the bone above the insertion of the internal pterygoid muscle: it is part of the deep cervical fascia projecting under the jaw.

The *stylo-maxillary ligament*, O, reaches from the styloid process to the hinder and lower parts of the ramus of the jaw: this piece of the cervical fascia is continuous below with that separating the parotid and submaxillary glands (Plate xvi., N).

DESCRIPTION OF PLATE XXII.

The dissection of the submaxillary region is indicated in this Figure.

The steps of the dissection are the following:—The soft parts over the jaw are to be divided, and the bone is to be sawn through rather on the right of the symphysis; then, the tongue having been drawn out of the mouth, the mucous membrane is to be cut along it below, to trace forwards the vessels and nerves.

To make tense the muscles, fasten down the os hyoides with a stitch to one of the firm surrounding parts.

MUSCLES OF THE TONGUE AND THE HYOID BONE.

Extrinsic muscles of the tongue and elevators of the os hyoides occupy the interval between the tongue and that bone.

A. Mylo-hyoideus, reflected.
B. Genio-hyoideus.
C. Genio-glossus.
D. Hyo-glossus.
E. Stylo-glossus.
F. Stylo-hyoideus.
G. Middle constrictor.
H. Digastricus.

J. Inferior constrictor.
K. Thyro-hyoideus.
L. Omo-hyoideus.
N. Sterno hyoideus.
O. Stylo-hyoid ligament.
P. Great cornu of the hyoid bone.
Q. Thyroid cartilage.

Elevators of the os hyoides. Some of the muscles of this group, viz., the mylo-hyoideus, A, the stylo-hyoideus, F, and the digastricus, H, have been described (p. 144): the remaining elevator is given below.

Genio-hyoideus, B. It arises from an eminence inside the symphysis of the jaw, and is inserted below into the centre of the body of the hyoid bone. The muscle touches its fellow along the middle line, and lies between the genio-glossus, C, and the mylo-hyoideus, A.

When the mouth is shut the muscle will raise the hyoid bone; or the

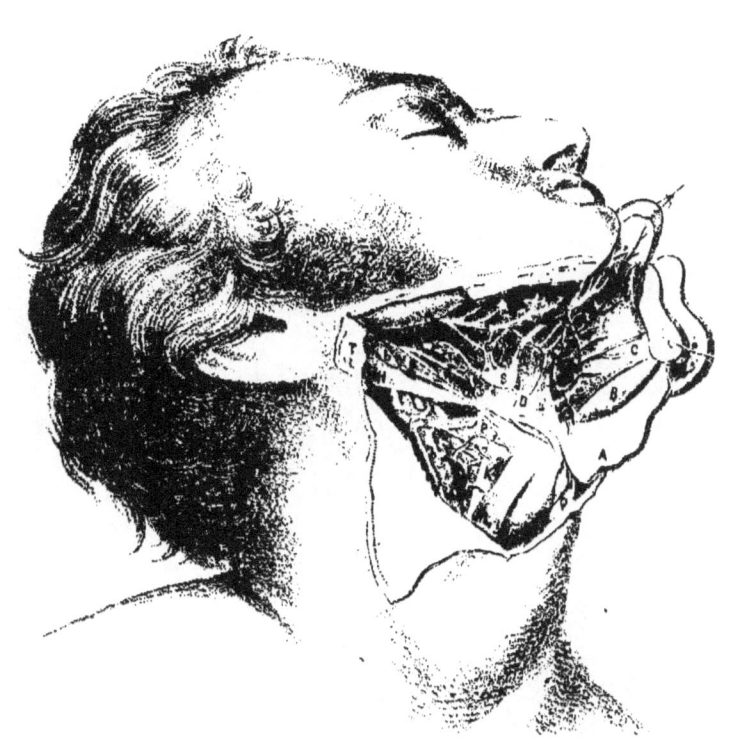

os hyoides being fixed, it will help to bring down the jaw, as in the act of opening the mouth.

Extrinsic tongue muscles. There are four on each side, viz., hyo-glossus, genio-glossus, stylo-glossus, and chondro-glossus: only the three first are now laid bare.

Hyo-glossus, D. This thin muscle arises from the hyoid bone, viz., from the great cornu by one part (cerato-glossus), and from the body of the hyoid bone by another * (basio-glossus). From this attachment the fibres ascend and enter the side of the tongue.

With the os hyoides fixed the hyo-glossus can depress the tongue in the floor of the mouth, and give to that organ a rounded form. Supposing the tongue the fixed point the muscle will raise the hyoid bone, preparatory to swallowing.

The *stylo-glossus*, E, arises from the styloid process and the stylo-maxillary ligament (Plate xx.), and enters the back of the tongue; but its fibres extend forwards underneath the side of the tongue to the tip where they blend with their fellows.

The muscles of opposite sides contracting will draw back and up the base of the tongue; and by the action of one muscle the point of the tongue will be turned to the same side of the mouth.

Genio-hyo-glossus, C. Shaped like a fan, it arises by a narrowed part from a tubercle inside the symphysis of the jaw; and it is inserted along the middle of the tongue from tip to root, as well as into the body of the hyoid bone. In contact with its fellow by the inner surface, the anterior edge is covered by the mucous membrane of the mouth, and the posterior touches the genio-hyoideus, B.

All the fibres contracting the tongue will be sunk in the floor of the mouth, and notably its middle part, so as to give a concavity to the upper surface. If only the lower fibres act they will raise the hyoid bone, and put forwards the tongue between the teeth: by means of the last mentioned fibres the muscle will be able to dilate the pharynx anteriorly.

The *stylo-hyoid ligament*, O, stretches between the end of the styloid process and the small cornu of the hyoid bone. Below, it lies beneath the hyo-glossus, and gives attachment to the middle constrictor, G.

---

* A third fleshy slip (chondro-glossus), which is attached to the small cornu of the bone, is considered to form part of the muscle.

Sometimes this band is large and cartilaginous or even osseous; at other times it is slight, and so membranous as not to be recognized.

The *Pharynx*. In front of the carotid bloodvessels is the upper dilated part of the gullet, or the pharynx. Its wall contains thin muscles which overlap one another, and the chief of these are called constrictors: two are marked with G and J, but they will be more fully noticed in Plate xxv.

SALIVARY GLANDS.

The sublingual gland and parts of the submaxillary and parotid, are exposed in the dissection.

R. Sublingual gland.
S. Piece of the submaxillary.

T. Part of the parotid.
† Wharton's duct.

Submaxillary gland, S. A deep part of the gland projects beneath the mylo-hyoid muscle, and with it the following excretory duct is connected:—

The duct of the gland, †, (Wharton's,) is about two inches long; it ascends beneath the gustatory nerve and the sublingual gland to the floor of the mouth, and ends in an eminence on the side of the frænum linguæ.

The *sublingual gland*, R, lies under the fore part of the tongue, where it forms a prominence, but it is separated from the cavity of the mouth by the mucous membrane. Elongated from before back, it is about one inch and a half in length, and meets its fellow in front.

Its ducts are numerous (8 to 20), and open for the most part by separate orifices in the floor of the mouth, but some join the duct of the submaxillary gland.

LINGUAL VESSELS.

The vessels of the tongue are few in number, in comparison with the nerves, there being but one on each side.

a. Common carotid trunk.
b. External carotid artery.
c. Upper thyroid branch.
d. Lingual artery.
e. Ranine branch.

f. Sublingual branch.
g. Facial artery, cut.
h. Occipital artery.
i. Branch of the sublingual artery.
l. Internal jugular vein.

The *lingual artery, d,* springing from the external carotid, runs obliquely upwards beneath the hyo-glossus to the under surface of the tongue, where it takes the name *ranine,* and continues along the middle line to the tip—distributing offsets. Near the front of the tongue the arteries of opposite sides correspond with the frænum linguæ in position, and may be cut when that fold of the mucous membrane is snipped with a scissors in tongue-tied infants.

A few named branches come from the artery: the most unimportant is the *hyoid* branch, which supplies one or more of the muscles attached to the os hyoides.

Beneath the hyo-glossus a *dorsal lingual* branch takes its origin. And at the fore part of that muscle arises the *sublingual* branch, *f,* which supplies the gland of the same name and the contiguous muscles, and joins the artery of the opposite side by means of the twig, *i.*

Lingual vein.—Its anatomy is similar to that of the artery, and it ends in the internal jugular vein.

NERVES OF THE TONGUE.

Six large nerves end in the tongue, three in each half; and the three of the right side are delineated as they course through the submaxillary region.

1. Glosso-pharyngeal nerve.
2. Hypoglossal nerve.
3. Descendens noni branch.
4. Upper laryngeal nerve.
5. Gustatory nerve.
6. Submaxillary ganglion.
7. Loop between the gustatory and hypoglossal nerves.

The *hypoglossal nerve,* 2 (twelfth cranial, Plate xxiv.), is the motor nerve of the tongue. Coursing with the lingual artery as far as the hyo-glossus it passes over this muscle, and enters the fibres of the genio-hyo-glossus, in which it is continued to the tip of the tongue, gradually decreasing in size by the supply of offsets.

On the hyo-glossus it furnishes branches to three extrinsic tongue muscles—the hyo, stylo, and genio-glossus; and to one elevator of the hyoid bone—genio-hyoideus. It joins the gustatory nerve, 5, by means of the loop, 7.

The *glosso-pharyngeal* nerve, 1 (ninth cranial), taking the course of the stylo-pharyngeus muscle (Plate xxiv.), enters beneath the hyo-glossus to reach the mucous membrane and the papillæ of the hinder third, and the lateral part of the tongue. Beneath the hyo-glossus muscle it furnishes offsets to the pharynx, the arches of the soft palate, and the tonsil.

The nerve confers sensibility on the mucous membrane of the pharynx, and gives the faculty of tasting in the back of the tongue and in the pillars of the soft palate.

The *gustatory nerve*, 5, coming from the pterygoid region (Plate xxi.) appears between the jaw and the internal pterygoid muscle, and courses forwards along the under surface of the tongue to the tip. At first the nerve rests against the prominence inside the last molar tooth; and in the rest of its extent in the tongue it lies near the edge, covered by the mucous membrane.

Offsets from it supply the mucous membrane of the floor of the mouth, the submaxillary and sublingual glands, and the tongue in front of the distribution of the glosso-pharyngeal nerve—especially the mucous membrane and the papillæ.

As this branch of the fifth cranial nerve does not receive any filaments of the motor root (p. 171) its function is sensory; and the faculty of tasting in the fore part of the tongue is dependent upon it.

Submaxillary ganglion, 7. This little body resembles the lenticular ganglion in the orbit (Plate xiv.), and is connected with the branch of the fifth nerve distributed to the tongue. Smaller than the lenticular ganglion, and occasionally reddish in color, it lies just above the submaxillary gland.

Some branches are furnished to the submaxillary gland and the mucous membrane of the mouth.

Other branches, sometimes called roots, join with the surrounding nerves, like the communicating branches of the lenticular ganglion. Thus the ganglion is connected above with the gustatory—a sensory nerve; with the facial—a motor nerve, by means of the chorda tympani (p. 171), which runs by the side of the gustatory to the tongue, and gives a slender thread to the back of the ganglion; and with the sympathetic through the plexus of that nerve on the facial artery.

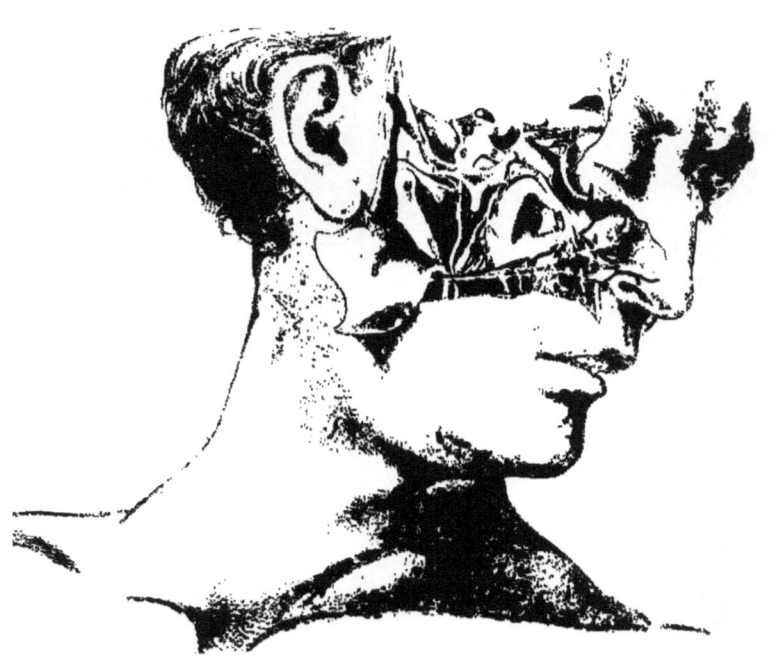

DESCRIPTION OF PLATE XXIII.

In this dissection the second trunk of the fifth nerve, and a part of the internal maxillary artery, are brought into view.

Supposing the head and the orbit opened, the dissection will be completed by removing the outer wall of the orbit, and the side of the cranium forming part of the middle fossa of the base of the skull.

SOME MUSCLES OF THE FACE.

Some of the muscles of the eyelids and upper lip being partly displayed will be referred to shortly; the other muscles, viz., those of mastication, have received sufficient notice already.

A. External pterygoid muscle.
B. Masseter muscle.
C. Buccinator muscle.
D. Levator anguli oris.
E. Levator labii superioris.
F. Levator lab. sup. alæque nasi.
G. Orbicularis palpebrarum.
H. Rectus oculi superior.
I. Antrum maxillare.
L. Oliquus oculi inferior.

Orbicularis palpebrarum, G. This thin sphincter muscle occupies the eyelids, forming loops around their aperture, and extends beyond the margin of the orbital cavity.

When the fibres contract the lids are closed, the upper one being specially brought down; and the integuments around the eye are wrinkled, and drawn towards the nose. In paralysis of the muscle the eyelids cannot be brought together, and the eyeball remains constantly uncovered.

Elevators of the upper lip. Three muscles raise the upper lip, viz., a common and a special elevator, and an elevator of the angle of the mouth.

The *elevator of the angle*, D, arises from the canine fossa of the upper jaw bone, and blends at the corner of the mouth with other muscles.

The *special elevator*, E, arises from the margin of the orbit over the infra-orbital foramen, and joins the sphincter of the mouth.

The *common elevator*, F, arises from the upper jaw bone at the inner side of the orbit, and ends at the mouth like the preceding: it is attached also to the wing of the nose by a separate slip.

These muscles contracting together will raise the upper lip; but the elevator of the angle can act independently of the others, and raise the corner of the mouth. Commonly, elevation of the lip and of the wing of the nose follows forced contraction of the sphincter of the eyelids, in consequence of a fleshy slip being prolonged from the orbicularis to the special elevator.

INTERNAL MAXILLARY ARTERY.

Two of the terminal branches of the internal maxillary artery at the spheno-maxillary fossa are traced out in the dissection.

- a. Internal maxillary artery.
- b. Posterior dental branch.
- c. Infra-orbital branch.
- d. Buccal branch.
- e. Internal carotid artery in the skull.
- f. Ophthalmic artery.
- g. External carotid trunk.
- h. Superficial temporal branch.
- n. Transverse facial branch.

The *posterior dental* artery, b, springing from the internal maxillary near the spheno-maxillary fossa, is inclined downwards and forwards over the upper maxilla to the front of the bone, and anastomoses with the infra-orbital artery.

It supplies superficial and deep branches:—the former descend to the buccinator muscle, the periosteum, and the gums; and the latter enter foramina in the bone, and supply offsets to the fangs of the teeth and to the lining membrane of the antrum maxillare.

The *infra-orbital* artery, c, arises near the preceding, and enters the infra-orbital canal with the upper maxillary nerve. Continued through that canal, it issues at the infra-orbital foramen, and ends in branches for the lower eyelid and the parts between the orbit and mouth: it communicates with the facial, transverse facial, and posterior dental arteries.

Many small offsets are furnished to the orbit; and near the front of the jaw bone it sends downwards an *anterior dental* branch, with a nerve, 8, of the same name, to supply the incisor and canine teeth.

The *transverse facial* artery, n, commonly a branch of the superficial temporal, crosses the side of the face, supplying the contiguous parts, and anastomoses with the facial and the infra-orbital arteries.

UPPER MAXILLARY NERVE.

The second trunk of the fifth cranial nerve (p. 107) is named as above from passing through the upper maxilla; it supplies the teeth of the upper jaw.

1. Posterior dental branch.
2. Upper maxillary nerve.
3. Optic nerve.
4. Orbital branch, cut.
5. Ophthalmic trunk.

6. Inferior maxillary trunk.
7. Buccal branch.
8. Anterior dental branch.
9. Branches of the facial nerve.

The *upper maxillary* nerve, 2, leaves the skull by the foramen rotundum, and courses to the face across the spheno-maxillary fossa, and through the infra-orbital canal. In the face it splits into large branches which are distributed to the muscles and the integuments between the eye and the mouth: a fine offset ascends with a small artery to the lower eyelid and the orbicular muscle. Its facial or terminal branches join in a plexiform manner with branches of the facial nerve. It gives off the following branches:—

Dental branches:—These are usually two in number, one at the back, and the other at the front of the maxilla.

The posterior branch, 1, descends on the jaw, gradually diminishing in size, and is received into a canal in the bone. Most of its branches course through the bone to supply the grinding teeth, but one or two slender offsets are furnished to the gums and the buccinator muscle.

The anterior branch, 8, is conducted by a bony canal in front of the antrum to the bicuspid and incisor teeth: it sends some filaments to the mucous lining of the nose, and joins the posterior branch.

Orbital and *spheno-palatine* branches:—Opposite the spheno-maxillary fossa these remaining branches take origin.

The orbital branch, 4 (temporo-malar), is a cutaneous nerve of the face and temple, and receives its designation from passing through the cavity of the orbit. In the dissection it was cut necessarily by the removal of the outer wall of the orbit. In its uninjured state the nerve can be traced into the orbit, where it splits into a temporal and a malar branch; these issue to their destination through apertures in the malar bone.

The spheno-palatine branches, two in number, descend beneath the internal maxillary artery, and communicating with Meckel's ganglion in the spheno-maxillary fossa, supply the lining membrane of the nose and roof of the mouth; the soft palate, and the tonsil; and the mucous lining of the pharynx near the aperture of the Eustachian tube.

The upper maxillary trunk of the fifth nerve springs from the Gasserian ganglion without commixture with the motor root, and is solely a sensory nerve, like the first or ophthalmic trunk. To its influence is due the sensibility of a part of the face, of the teeth of the upper jaw, of the nose cavity, and of the roof of the mouth and the soft palate.

Facial nerve.—This branch, of rather large size, which is marked with 9, is called infra-orbital: it lies below the orbit, and supplies the muscles between the eye and mouth, and on the nose. In its course inwards its offsets cross, and join the branches of the upper maxillary nerve, forming the infra-orbital plexus by this arrangement.

The facial is the motor nerve chiefly of the muscles of the face and head; and it is distributed for the most part to muscles receiving sensibility from the three trunks of the fifth cranial nerve. To the buccinator, which acts as a muscle of mastication as well as a dilator of the aperture of the mouth, it gives the ability to contract; and consequently this muscle is paralyzed when the other muscles which are supplied by the facial nerve lose their contractile power.

DESCRIPTION OF PLATE XXIV.

This Illustration will serve as a guide to the dissection of the internal carotid and ascending pharyngeal arteries, and of the cranial nerves distributed in the neck.

After the examination of the pterygoid region and the upper maxillary nerve, the dissection delineated will be prepared by detaching the styloid process with its muscles, and the external carotid artery and its branches; and by sawing off the large piece of the side of the skull out-

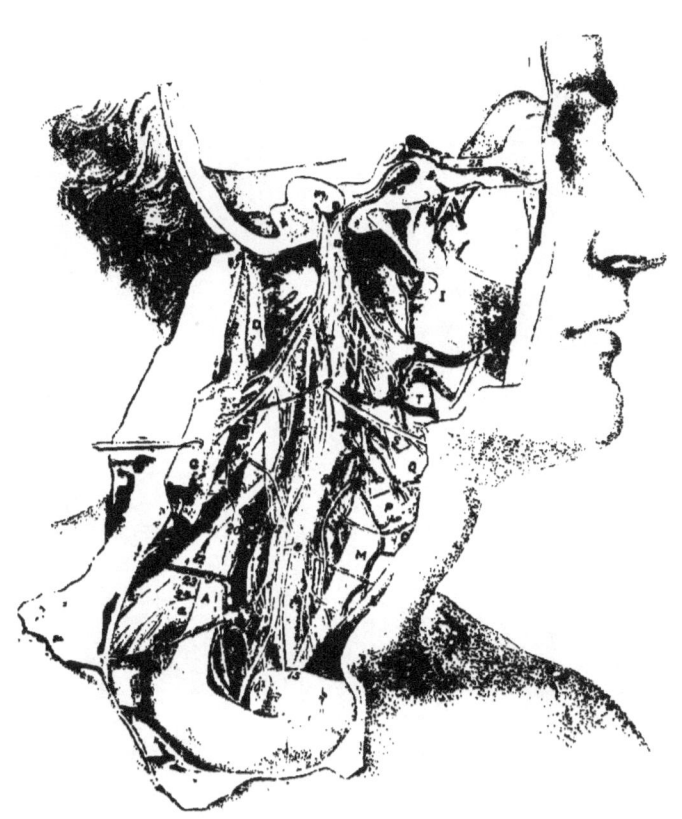

side the jugular foramen and the carotid canal. Finally a dense fibrous tissue surrounding the nerves and vessels near the base of the skull should be taken away carefully; and as the internal jugular vein renders obscure the view of many objects it may be removed.

DEEP MUSCLES OF THE NECK.

Lying on the front of the spinal column are the deep muscles for the flexion and rotation of this part of the spine, and of the head, which will be now described. And superficial to the level of the carotid bloodvessels is the group of muscles, before referred to in part, which belongs to the pharynx and tongue, and the hyoid bone and the larynx.

A. Scalenus anticus.
B. Scalenus medius.
C. Sterno-mastoideus.
D. Splenius capitis, cut.
E. Levator anguli scapulæ.
F. Obliquus capitis inferior.
G. Obliquus capitis superior.
H. Rectus capitis lateralis.
I. Pterygo-maxillary ligament.
J. Rectus capitis anticus major.
K. Longus colli.
L. Sterno-hyoideus.
M. Sterno-thyroideus.
N. Thyroid body.
O. Omo-hyoideus.

P. Thyro-hyoideus.
Q. Hyo-glossus.
R. Constrictor inferior.
S. Constrictor medius.
T. Submaxillary gland.
U. Styloid process, cut off.
V. Stylo-glossus.
W. Stylo-pharyngeus.
X. Constrictor superior.
X'. Buccinator.
Y. Levator palati.
Z. Tensor palati.
‡ Cartilage of the Eustachian tube.

The *rectus capitis lateralis*, H, resembles in position an inter-transverse muscle; it is attached below to the lateral part of the atlas, and above to the jugular eminence of the occipital bone.

The muscle will approximate the skull to the atlas, and so help in inclining the head towards the shoulder.

The *rectus capitis anticus major*, J, is continued upwards in a line with the anterior scalenus. Arising below where the scalenus is attached, viz., from the transverse processes of the 6th, 5th, 4th, and 3d cervical vertebræ, it ascends, becoming thicker near the skull, to be inserted into the basilar process of the occipital bone.

If the muscles of opposite sides act the head will be bowed forwards;

but only one acting it will turn the face to its own side, in consequence of its oblique position.

A third rectus muscle (rect. cap. anticus minor) lies between the two preceding, and passes from the atlas to the basilar process of the occipital bone; it is concealed by the vessels and nerves near the head.

The *longus colli* muscle, K, lies on the front of the spinal column, between the atlas and the second dorsal vertebra; and it is attached to the bodies of the vertebræ and to certain of the transverse processes. For the complete display of the muscle the pharynx should be detached.

The muscle bends forwards the spine, and can rotate the same by means of the lateral slips connected with the transverse processes of the vertebræ.

SUBCLAVIAN AND CAROTID ARTERIES.

In this Illustration the internal carotid artery can be observed throughout; and by means of the Figure a more complete view of some of the branches of the subclavian and carotid trunks may be obtained.

a. Subclavian trunk.
b. Vertebral artery.
c. Internal mammary branch.
d. Thyroid axis.
e. Inferior thyroid artery.
f. Supra-scapular artery.
g. Transverse cervical artery.
h. Ascending cervical branch.
i. Common carotid trunk.
k. Upper thyroid artery.
l. Crico-thyroid branch.
m. External carotid trunk.

n. Laryngeal branch.
o. Lingual artery.
p. Facial artery.
q. Occipital artery, cut; with a branch to sterno-mastoideus.
r. Tonsillitic branch.
s. Inferior palatine branch.
t. Ascending pharyngeal artery.
u. Internal carotid trunk.
x. Internal maxillary artery, cut.
y. Internal jugular vein, cut.

Subclavian trunk.—The arch of the subclavian artery, and the branches of its first part (p. 154) are here represented.

In this body a rare condition of the *inferior thyroid artery, e,* existed;* the vessel sprang from the vertebral artery, *b,* instead of the thyroid axis, *d,* and then took its usual course to the thyroid body, N.

---

* In Mr. Quain's Surgical Anatomy of the Arteries, p. 169, it is said to have been seen once.

Common carotid artery, i.—The extent and situation of the artery, and the nerves in connection with it (p. 146) can be well perceived in this Plate. In this body the arterial trunk splits into two at a point higher than usual.

External carotid trunk, m.—Only the lower part of the artery, with its first branches which were not represented or only imperfectly in preceding Plates, has been left in the dissection. And as the carotid begins above the usual place these first branches have to descend to their destined positions.

The *upper thyroid, k*, runs over the superficial surface of the thyroid body, N, before entering the substance. It furnishes, firstly, muscular offsets; next a laryngeal branch, *n ;* and lastly, a crico-thyroid branch, *l*, which lies on the membrane of the same name, joining that of the opposite side, and would be endangered in the operation of laryngotomy.

Lingual artery, o :—its hyoidean offset arises before the artery passes beneath the hyo-glossus, Q, and is distributed to the thyro-hyoideus, P.

The *facial artery, p*, furnishes the following branches to the neck before it reaches the jaw:—

A tonsillitic offset, *r*, ascends between the pterygoideus internus and the stylo-glossus, V, and perforating the upper constrictor, X, ends in the tonsil and the side of the tongue.

An inferior palatine branch, *s*, courses along the side of the pharynx between the stylo-glossus and stylo-pharyngeus muscles to the upper border of the superior constrictor, X, where it passes inwards to supply the palate. It supplies muscular branches; and one offset, long and slender, reaches the Eustachian tube ‡.

Other offsets of the facial, viz., submental and glandular are seen in Plate xvii.

The *ascending pharyngeal artery*, arises near the beginning of the external carotid, and ascends on the spinal column between the pharynx and the internal carotid trunk nearly to the skull. Here it enters the pharynx above the upper constrictor, and ends in branches to the front and back of the soft palate; of these the anterior are the largest, and join with corresponding branches of the opposite side, so as to form two arches beneath the mucous membrane—one lying near the upper, and the other near the lower edge of the velum palati (Quain).*

---

* Fifth edition of Quain's Anatomy, 1846, p. 489.

Branches are given to the contiguous muscles, the lymphatic glands, and the nerves; and one (meningeal) enters the skull through the foramen lacerum, and ends in the dura mater.

The *internal carotid* artery, *u*, ascends through the neck and the temporal bone to the interior of the cranium, and terminates in branches for the brain and the orbit.

The cervical part of the vessel, of the same size throughout and devoid of branches, lies by the side of the pharynx, and rests on the rectus anticus, J. At first the artery is accessible in an operation (p. 148), but it becomes deep afterwards beneath the parotid gland and the digastricus, and the styloid process and its muscles.

The internal jugular vein is contained in a sheath of fascia with the artery, and is external or posterior to it.

Numerous nerves are in contact with the vessel. Crossing it superficially from above down are the glosso-pharyngeal, 1, the pharyngeal branch of the vagus, 5, and the hypoglossal nerve, 7; and beneath it, also with a cross direction, are the pharyngeal branches of the sympathetic, the upper laryngeal, 3, and the external laryngeal, 4. In the sheath between it and the vein, and parallel to it, lies the vagus nerve, 2; and behind the sheath and parallel, is placed the sympathetic nerve with its branches. Close to the skull the cranial nerves issuing by the foramen jugulare and anterior condyloid foramen interpose between the artery and vein, but they diverge afterwards to their destination.

In the temporal bone the artery becomes flexuous, and fills the carotid canal, only a few branches of the sympathetic ascending around it: here it gives a small tympanic branch to the ear, which pierces the bone.

For the anatomy of the artery in the skull see Plate xiii. (p. 109); and for the description of the ophthalmic artery refer to p. 113.

DEEP NERVES OF THE NECK.

Four cranial nerves, and the sympathetic nerve, with their branches, together with the spinal nerves of the neck, are visible in the Plate.

1. Glosso-pharyngeal nerve.
2. Vagus nerve.
3. Upper laryngeal nerve.
4. External laryngeal nerve.
5. Pharyngeal branch.
6. Spinal accessory nerve.
7. Hypoglossal nerve.
8. Descendens noni branch.

DEEP NERVES OF THE NECK. 185

9. Communicating branch from the spinal nerves.
10. Recurrent laryngeal nerve.
11. Cord of the sympathetic nerve.
12. Upper cervical ganglion.
13. Middle cervical ganglion.
14. Lower cervical ganglion.
15. Middle cardiac nerve.
16. First cervical nerve (loop of the atlas).
17. Second cervical nerve.
18. Third cervical nerve.
19. Fourth cervical nerve.
20. Phrenic nerve.
21. Fifth cervical nerve.
22. Sixth cervical nerve.
23. Seventh cervical nerve.
24. Eighth cervical nerve.
25. Supra-scapular nerve.
26. Carotid branches of the sympathetic.
27. Upper maxillary nerve.
28. Optic nerve.
†† Cardiac branches of the vagus in the neck.

The *glosso-pharyngeal* or ninth cranial nerve, 1, leaves the skull by the jugular foramen, and courses to the pharynx over the carotid artery; passing then beneath the hyo-glossus muscle, Q, it ends in terminal branches for the tongue. In the foramen of exit the nerve possesses two small ganglia, and furnishes a branch (Jacobson's nerve) to the tympanum. Its branches beyond the cranium are the following:—

As it crosses the carotid artery some fine filaments descend on the vessel, and join the sympathetic and the pharyngeal branch, 5, of the vagus.

Muscular branches enter the stylo-pharyngeus and the upper two constrictors; and at the side of the pharynx it joins in a plexus (pharyngeal) with offsets of the sympathetic and of the pharyngeal branch of the vagus.

Numerous offsets are distributed to the mucous membrane of the pharynx opposite the opening of the mouth.

The nerve is chiefly sensory in its function, and it confers on a part of the tongue the faculty of tasting as before said (p. 176); but as the stylo-pharyngeus muscle is supplied altogether by it some motor influence must be obtained from it. By means of its branches to the lining of the pharynx impressions produced by the presence of food are conveyed to the sensorium.

The *pneumo-gastric*, vagus, or tenth cranial nerve, 2, issues from the skull by the foramen jugulare. In the aperture of exit it has a ganglion (gang. of the root); and it gives a branch to the ear, like the glosso-pharyngeal.

Beyond the skull it is continued through the neck to the thorax, lying in the carotid sheath between the artery and the jugular vein; and as it leaves the neck on the right side it crosses the subclavian artery. Near the skull it is marked by a long fusiform ganglion (gang. of the trunk), which is united with the hypoglossal nerve, 7. In the neck the nerve supplies the undermentioned branches to the pharynx, the larynx, and the heart.

The *pharyngeal* branch, 5, begins in the ganglion, and crosses over (sometimes under) the internal carotid, to reach the pharynx. After being joined by offsets of the glosso-pharyngeal, it communicates with the sympathetic and the superior laryngeal to form the pharyngeal plexus: it ends in the constrictor muscles.

The *upper laryngeal* nerve, 3, arises also from the ganglion, and courses under the carotid to the interval between the hyoid bone and the thyroid cartilage: here it pierces with an artery the thyro-hyoid ligament, and is distributed to the mucous membrane of the larynx. See Plate of the larynx.

Under the carotid it joins largely with the sympathetic nerve; and it furnishes the *external laryngeal* nerve, 4, which supplies the inferior constrictor, and ends in the crico-thyroideus muscle (Plate xxv.).

Cardiac branches, † †. One springs from the nerve trunk at the lower, and one or two at the upper part of the neck: they join branches of the sympathetic. In this dissection the upper communicated with the descendens noni nerve.

Recurrent or *inferior laryngeal* nerve, 10. On the right side this nerve arises as the vagus enters the thorax, and winding round the subclavian artery, runs back to the larynx: it is distributed chiefly to the laryngeal muscles. See Plate xxvii. On the left side the nerve begins in the thorax opposite the arch of the aorta, round which it makes a loop to come back to the larynx.

In the neck the pneumo-gastric nerve ramifies in the walls of the air and food passages, and bestows sensibility on the mucous membrane and contractility on the muscular structure; but the contraction of the muscles supplied not being placed under the control of the will (except those of voice), the nerve resembles more the sympathetic than the other motor cranial nerves.

From the partial mixing of its motor and sensory nerve fibres the branches in the neck have different functions. Experiments seem to determine that the pharyngeal branch is a motor nerve; the superior laryngeal, chiefly sensory; and the recurrent laryngeal a motor nerve of the muscles of the larynx, but involuntary motory, and sensory to the muscular fibres in the trachea. The small cardiac branches are probably involuntary motory, and sensory in function like those to the lung.

The *spinal accessory* or eleventh cranial nerve, 6, comes out of the skull by the foramen jugulare, and communicates in that aperture with the vagus by means of an accessory piece.

Beyond the foramen the nerve is directed downwards and backwards to the sterno-mastoideus, which it pierces, and to the Trapezius muscle (Plate xv. p. 134). It joins freely with branches of the cervical plexus, and supplies with them the two muscles named.

This nerve resembles a spinal nerve in arising from the spinal cord, and in being moto-sensory in function; and this double function is not altogether dependent upon its union with the spinal nerves, for it alone may supply the sterno-mastoideus.

The *hypoglossal*, or twelfth cranial nerve, 7, leaves the skull by the anterior condyloid foramen, and turning over the vagus, with which it is inseparably united, descends as low as the digastric muscle before it is directed forwards to the tongue. No offset is distributed from the first part of the nerve, though it joins the vagus, the sympathetic, and the first spinal nerve; but many muscular branches arise from the last part of the hypoglossal, as may be seen in Plate xxiii.

It is supposed to be altogether a motor nerve at its origin; and it is thought that any sensory influence possessed by it is derived from its junction with other nerves near the skull.

Sympathetic nerve.—The cervical part of the sympathetic nerve, 11, lies on the spine beneath the great bloodvessels, and is continuous with the knotted cord in the thorax. In the neck it is marked by three ganglia—upper, middle and lower; and each ganglion furnishes external or communicating branches, internal or visceral, and branches to bloodvessels.

The *upper ganglion,* 12, is the largest of the three: it is fusiform in shape, with a reddish color, and is about two inches long. Near the base of the skull the cranial nerves lie over it.

The outer branches communicate with the first four spinal nerves, and with the tenth and twelfth cranial nerves.

Most of the inner branches pass beneath the carotid to join in the pharyngeal plexus; but one, larger than the rest and named *upper cardiac*, descends beneath the artery to the cardiac plexus in the thorax.

The nerves to bloodvessels from the *ganglion* (nervi molles) ramify on both carotid arteries, forming plexuses on them; and on some of the branches of the external carotid there are interspersed ganglia. Through the offset, 26, on the internal carotid the vessels and the vascular membrane of the brain are supplied, and communications take place with the cranial nerves in the middle fossa of the base of the skull.

The *middle ganglion*, 13, variable in size and shape, is placed near the inferior thyroid artery, *e*, and is smaller than the others. Its offsets are the following:—

Outer branches which join usually the fifth and sixth spinal nerves.

Inner branches ramify on the thyroid artery and end in the thyroid body. One of these, the *middle cardiac* nerve, 15, is continued to the cardiac plexus in the thorax.

The *inferior ganglion* lies beneath the subclavian artery and close above the neck of the first rib. It is rather rounded in shape, and is often divided into parts, as in the Figure, where one of the pieces is marked, 14. Its branches are similar to those of the other ganglia.

Outer branches, two or more in number, join the two lowest cervical nerves.

One large inner or visceral branch, *inferior cardiac*, runs beneath the subclavian artery to the cardiac plexus in the thorax.

Offsets to the bloodvessels entwine around the vertebral artery, *b*, forming a plexus on it; and other nerves ramify on the subclavian trunk which they surround with one or two loops.

The branches of the sympathetic in the neck serve chiefly to connect this nerve with others, and to supply the bloodvessels.

By means of the communicating branches with the cranial and spinal nerves the sympathetic gives fibres to, and receives fibres from those nerves; and the offsets joining the anterior primary trunk of each spinal nerve are directed through the roots of the nerve towards the spinal cord, and send also some fibres to the trunk of the nerve to be distributed peripherally with it.

To the bloodvessels the sympathetic gives the power of regulating the quantity of blood circulating through them; so that on section of its nerves (vaso-motory) to them the muscular coat is paralyzed, and being unable to contract on the contained fluid, the blood slackens in speed, and congestion of the vessels of the part and increased heat ensue. Stimulating the cut nerves by galvanism will restore for the time contraction of the muscular coat, and will cause a decrease in the congestion and the heat.

Spinal nerves.—Eight in number, they are divided equally between two plexuses;—the upper four entering the cervical, and the lower four the brachial plexus.

Cervical plexus.—The anterior primary branches of the first four nerves interlace in the cervical plexus: they are marked 16 to 19 inclusive, and the small branch of the first, 16, is named the loop of the atlas. The superficial offsets of the plexus are delineated in Plate xv.; the deep branches follow below:—

Branches to muscles.—From the loop between the first two nerves branches are furnished to the contiguous recti muscles; and from the other loops of the plexus the surrounding muscles, viz., the sterno-mastoideus C, Levator anguli scapulæ E, scalenus medius B, intertransversales, trapezius, and the platysma, receive nerves.

Communicating branches.—Offsets unite the loop of the atlas with the vagus and hypoglossal nerves, 2 and 7, and with the upper ganglion of the sympathetic, 12. And two small branches from the second and third nerves, 17 and 18 (in this case one comes also from the fourth nerve, 19) join in one, 9, which unites with the descendens noni, and assists to supply the depressor muscles of the hyoid bone.

The *diaphragmatic* or *phrenic* nerve, 20, begins in the fourth cervical nerve, but it often joins the trunk of the fifth nerve, 21, as it passes by. It descends to the thorax over the scalenus anticus A, and inside the internal mammary artery, *c*, as in the Drawing.

Brachial plexus.—The lower four nerves are much larger than the upper, and are prolonged to the upper limb. The trunks, marked from 21 to 24 inclusive, issue between the anterior and the middle scalenus,* and join with part of the first dorsal in the large cords seen in the Figure.

Only two of the branches arising from the plexus above the clavicle are now visible, viz., the supra-scapular 25, and the small nerve to the subclavius; the rest of this set of branches are shown in Plate xv. (p. 134).

---

* In this body the fifth and the fourth cervical came in front of the anterior scalenus.

DESCRIPTION OF PLATE XXV.

A SIDE view of the pharynx with its muscles is depicted in this Figure.

For this dissection the base of the skull is to be cut through behind the attachment of the pharynx; and the fore part of the head being fixed on a block, and the pharynx distended with tow, the muscles will be easily prepared.

THE PHARYNX AND ITS MUSCLES.

The pharynx is the upper part of the alimentary tube which is placed behind the nose, mouth, and larynx. Both the food and air pass along it. It reaches from the skull to the lower end of the larynx, gradually tapering from above down, and measures from five to six inches in length. Above it is inserted into the skull by a thin fibrous membrane called the aponeurosis of attachment of the pharynx; and in front it is fixed to the head, the hyoid bone, and the larynx.

In the wall of the pharynx are contained constricting and elevating muscles, which are employed in swallowing; the latter are engaged in placing the receiving bag in the position required for the entrance of the food or drink, and the former urge onwards to the gullet the morsel received.

A. Inferior constrictor.
B. Middle constrictor.
C. Superior constrictor.
D. Stylo-pharyngeus.
E. Levator palati.
F. Tensor palati.
G. Buccinator.
H. Stylo-glossus, cut.
I. Temporal muscle.
J. Mylo-hyoideus.
K. Sterno-hyoideus.
L. Omo-hyoideus.
M. Hyo-glossus.
N. Thyro-hyoideus.
O. Stylo-hyoid ligament, ossified.
P. Sterno-thyroideus.
Q. Crico-thyroideus.
R. Thyroid body, thrown down.
S. Œsophagus or gullet.
T. Trachea or windpipe.
† Pterygo-maxillary ligament.

The *constrictor muscles* are flat and thin, and are three in number on each side, viz., lower, middle, and upper. They are attached in front to the larynx, hyoid bone, and the head, and meet their fellows in the middle line behind: their contiguous edges overlap like scales, the upper being more superficial.

The *lower constrictor*, A (laryngo-pharyngeus), arises from the side of the cricoid and thyroid cartilages of the larynx; and the fibres end in the middle line behind. Its upper edge overlays the middle constrictor B, and the lower is continuous with the circular fibres of the œsophagus.

The *middle constrictor*, B (hyo-pharyngeus), is connected in front with the hyoid bone, viz. with the great and small cornua, and with the lower end of the stylo-hyoid ligament, O. The fibres radiate to their ending at the middle line behind, the lower passing beneath the inferior constrictor, and the upper over the superior constrictor to within an inch of the skull.

The *upper constrictor*, C (cephalo-pharyngeus), is fixed by its anterior edge to the following parts: to the pterygoid plate (lower third of the inner surface) and the hamular process, to the pterygo-maxillary ligament, †, to the lower jaw behind the last molar tooth, to the mucous membrane of the floor of the mouth, and to the side of the tongue. As the fibres pass back to the middle line, the upper form a free curved border below the skull, where the levator palati muscle E enters above them, and the lower are continued beneath the middle constrictor, and blend with fibres of the stylo-pharyngeus.

When these muscles contract, they diminish the size of the pharyngeal cavity by moving forwards the loose hinder part. In swallowing, the two lowest grasp and convey onwards by successive rapid contractions the morsel of food or the drink; whilst the upper one, which is placed above the aperture of the mouth, takes little share in the process, farther than by lessening the space above the mouth, it so far assists in opposing the ascent of the food behind the soft palate. As the tonsil is covered by the upper constrictor opposite the angle of the lower jaw, it may be compressed during the action of that muscle.

Elevators of the pharynx. Two muscles on each side, an external and internal elevator, descend from the head to raise the upper part of the pharynx preparatory to swallowing.

The *levator pharyngis externus*, D (stylo-pharyngeus), arises from the root of the styloid process, and descends, becoming wider, between the upper and middle constrictors to be inserted mainly into the upper border of the thyroid cartilage, and in part with the upper constrictor muscle.

Levator pharyngis internus (salpingo-pharyngeus) is delineated in Plate xxvi. N. It is a small muscular slip inside the pharynx, immediately beneath the mucous membrane, which arises by tendon from the end of the Eustachian tube, O, and joins below the palato-pharyngeus muscle, C.

The elevators make ready the pharynx for receiving the aliment, and they act in this way:—The large elevator draws upwards and outwards the part of the pharynx above the os hyoides, especially the part opposite the opening of the mouth; and elevates the larynx at the same time. And the small or internal elevator raises that part of the pharynx above the large elevator, which would become loose by the action of the other muscle.

Before deglutition takes place the hyoid bone is drawn forwards and upwards by its elevator muscles, giving thus increased size to the pharynx from before back; and the larynx is carried upwards and forwards at the same time under the tongue, so as to allow the opening into the windpipe to be placed in the position most favorable for its closure during the act of swallowing.

LARYNGEAL VESSELS.

Two arteries on each side supply the larynx, and the pharynx and windpipe in part.

a. Inferior thyroid artery.
b. Laryngeal branch.
c. Thyroid branch.
d. Laryngeal branch of the upper thyroid.
e. Lingual artery.
f. Internal carotid.

The *upper laryngeal* branch, d, is an offset of the superior thyroid artery, and enters the larynx through the thyro-hyoid membrane: its distribution in the larynx can be traced in Plate xxvii.

The *inferior thyroid* artery, a, ramifies by the branch, c, on the under part of the thyroid body; and sends a branch, b, into the interior of the larynx, which is delineated with the other laryngeal artery.

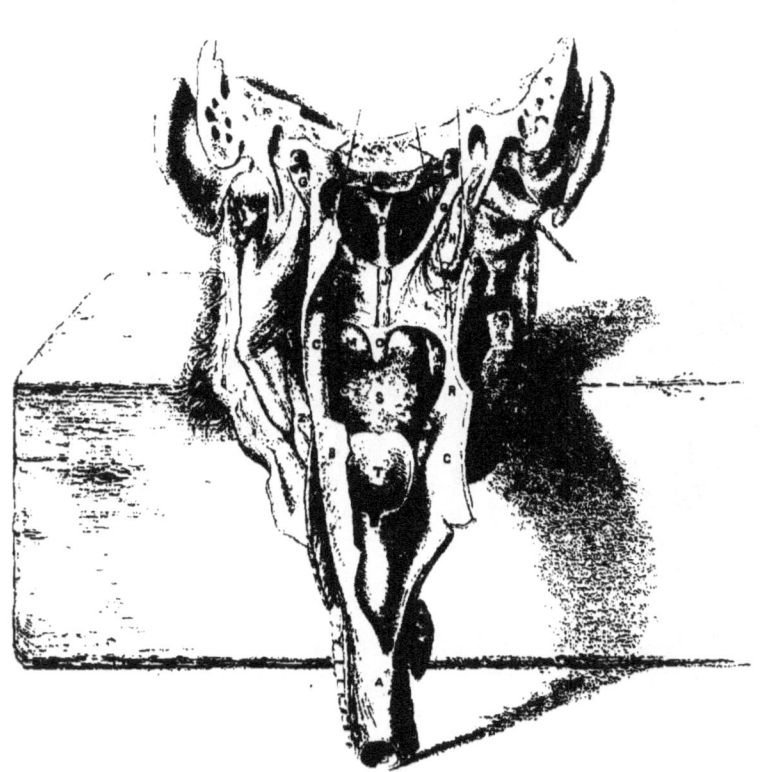

NERVES OF THE LARYNX.

Three of the nerves now apparent belong to the larynx and its muscles, and the remaining three enter the tongue.

1. Glosso-pharyngeal nerve.
2. Gustatory nerve.
3. Hypoglossal nerve.

4. Upper laryngeal nerve.
6. External laryngeal nerve.
7. Recurrent laryngeal nerve.

The *upper laryngeal* nerve, 4, enters the larynx through the thyro-hyoid membrane with the artery, and ends in the mucous membrane.

The external laryngeal branch, 6, arises from the preceding high in the neck, and is distributed outside the larynx to the crico-thyroid muscle, Q, and to the inferior constrictor A; and as it is the only nerve reaching that laryngeal muscle, it must give to the fibres sensibility and contractility.

The *inferior laryngeal* or recurrent nerve, 7, a branch of the vagus, ascends between the gullet and the windpipe, and passes under the inferior constrictor to supply the muscles of the larynx (Plate xxvii.). Muscular offsets are furnished by it to the two tubes between which it lies.

DESCRIPTION OF PLATE XXVI.

The interior of the pharynx, and the dissection of the muscles of the soft palate, are comprised in this Illustration.

The objects inside the pharynx will appear on slitting down the tube behind, and everting the edges: and the muscles of the soft palate will be laid bare by removing the mucous membrane on the left side, and that layer with some muscular fibres under it on the right, in the manner indicated.

INTERIOR OF THE PHARYNX.

The pharyngeal cavity reaches from the base of the skull to the lower edge of the cricoid cartilage of the larynx, and tapers from above down. At its middle it serves as a common passage for the air and food, but the upper part transmits air exclusively, and the lower part conveys only food. These three portions, differing thus in their use, have the following limits:—the upper reaches as low as the opening of the mouth, M, and communicates with the cavities of the nose and tympanum; the middle region extends from the mouth to the aperture of the larynx V; and the third portion lies beyond the larynx, and is continuous below with the œsophagus W. Along the front of the pharynx are seven openings.

A. Tube of the œsophagus.
B. Pharynx cut, and reflected.
C. Inner part of pharynx covered by mucous membrane.
D. Septum nasi.
E. Lower spongy bone.
F. Eustachian tube.
I. Buccinator muscle.
K. Soft palate.
M. Roof of the mouth.
N. Salpingo-pharyngeus muscle.
O. The uvula.
P. Anterior pillar of the palate.
Q. The tonsil.
R. Posterior pillar of the palate.
S. The tongue.
T. The epiglottis.
V. Upper opening of the larynx.
W. Opening of the œsophagus.
X. Internal pterygoid muscle.
Z. Mylo-hyoid muscle.

The *Eustachian tube*, F, one on each side, lies close to the base of the skull; on the right side the mucous membrane has been removed from the lower end. Its extremity in the pharynx is cartilaginous and membranous, and is dilatable; but the upper part is osseous, and is contained in the temporal bone. At its lower end the cartilage is enlarged, but more at the inner than the outer side, and gives to the tube a funnel-shaped opening. The pharyngeal aperture is oval from before back, and is placed close behind the internal pterygoid plate, to which the tube is united by fibrous tissue higher up; it is on a level with the inferior meatus,—the upper part of the opening reaching as high as the upper border of the lower spongy bone.

This tube leads from the pharynx to the middle ear or tympanum; it transmits air to the ear cavity, and allows the mucus of that space to escape through it. Ordinarily the lower end is closed, and the air is shut in the tympanum, but the pharyngeal opening can be rendered patent by the action of the palate muscles, so as to permit the passage of air. An instrument can be passed into it through the nose for the purpose of removing obstruction in the tube, or of conveying air into the tympanum.

The *posterior nares* are the apertures of communication between the two sides of the nose cavity and the pharynx. Each is elongated from above down, and will admit readily the tip of the finger. In the dried skull it is bounded by the vomer internally and the internal pterygoid plate externally, and by the body of the sphenoid above and the palate bone below; but in the fresh state the bones are clothed by the mucous membrane, though without much diminution in the size of the opening. Separating the two is the septum nasi, D.

These apertures allow the air to pass in and out when the mouth is closed. Each is very much larger than the opening in the face of the same side of the nasal cavity; and its increased size will be of use in communicating with the upper part of the nose, and in allowing the outgoing air to ascend towards the roof of the space, and warm the parts that have been rendered cooler in inspiration.

When the lower jaw is immovably fixed, liquid food can be passed into the stomach by a small flexible tube introduced into the pharynx through the nose and the posterior naris.

In hæmorrhage from the half of the nose the fluid may escape by the nostril, or the posterior naris, or by both those openings when the flow of blood is great, and it may be needful to check the loss of blood by stopping both openings. The aperture in the face can be closed easily; but the posterior naris will have to be plugged through the mouth.

The posterior opening of the mouth, M, is named *isthmus faucium*, and has the following bounds:—Below lies the tongue, S; and above are the soft palate, K, and the uvula. On each side is placed the anterior arch of the palate, P, consisting of a fold of mucous membrane with fibres of the palato-glossus muscle: these folds of opposite sides constitute the pillars of the fauces.

This opening marks the boundary line between the mouth and the pharynx, and all voluntary control over the morsel to be swallowed ceases at that spot. The anterior palatine arches on the sides of the aperture take part in the process of deglutition in this way:—as soon as the food has been moved backwards by the tongue to the isthmus, the lateral arches are shortened and moved inwards by the contraction of their contained muscular fibres, and shut off with the tongue the cavity of the mouth.

Upper aperture of the larynx, V.—This is a single opening, and occupies the middle line just below the mouth. Wide before and narrow behind it is sloped down and back; it extends upwards rather above the hyoid bone, and downwards to the bottom of the central notch in the front of the thyroid cartilage. In front it is bounded by the wide expanded part of the epiglottis, T; and behind by the tips of the cornicula laryngis, and by the arytænoideus muscle and the mucous membrane. Laterally it is limited by a fold of mucous membrane (arytæno-epiglottidean) which stretches from the epiglottis to the arytænoid cartilage, and contains the depressor muscle of the epiglottis.

Through this hole the air is inspired and expired in breathing; and during the respiratory act the space remains open with the epiglottis raised.

When deglutition is about to take place the larynx is moved upwards and forwards under the hyoid bone and the tongue, and the epiglottis is partly lowered; and during swallowing the epiglottis is placed over the orifice, so as to close it from the passing food or drink, whilst the muscular fibres on the sides and back of the opening contract, and give increased security against the entrance of the aliment into the windpipe. Even when the epiglottis is absent the food does not find its way into the air passage, because the upper part is sufficiently closed by the elevation of the larynx, and by the contraction of the muscular fibres around the upper opening and on each side of the passage lower down. If an attempt is made to take breath during, or too soon after a long draught, some of the fluid is drawn with the air under the partially-raised valve, and produces violent coughing by irritation of the larynx.

The *aperture of the œsophagus*, W, terminates inferiorly the cavity of the pharynx, and is placed opposite the lower edge of the cricoid cartilage: it is circular in form, and is surrounded by the fibres of the lower constrictor.

THE SOFT PALATE AND THE TONSIL.

The *soft palate* (velum pendulum palati) forms the loose and movable part of the roof of the mouth, and depends between the nose and mouth cavities. In a state of rest it hangs like a curtain behind the mouth; but it can be moved backwards by muscles to the wall of the pharynx, so as to act like a valve in separating the upper from the middle region of the pharynx.

It is attached above by an aponeurosis to the back of the hard palate; and it is constructed chiefly of muscles covered by mucous membrane. Laterally it is blended with the sides of the pharynx. At the lower edge it is free; and from its centre hangs a rounded elongated part, the uvula, O; whilst on each side two folds, the arches of the soft palate, are continued downwards from it.

The *arches* of the half of the *soft palate*, P and R, begin above, near the middle of the velum, and descend on the sides of the tonsil, Q, diverging from each other. The anterior, P, is continued in front of the tonsil to the side of the tongue near the base; and the posterior is directed behind the tonsil to the back of the pharynx. Each consists of a fold of mucous membrane inclosing muscular fibres: in the anterior fold is the palato-glossus muscle, and in the posterior lies the palato-pharyngeus.

Tonsil, Q. This body is an aggregate of ten to twenty follicular glands, like those over the root of the tongue (Kölliker), and it occupies the interval between the arches of the palate. Its size varies much. Its situation is marked by the presence of small holes in the mucous membrane, without any surface-prominence; but when enlarged from disease it projects, diminishing thus the size of the isthmus of the fauces, and forms a swelling which may be felt externally near the angle of the jaw.

In its structure it resembles the follicular glands. In the bottom of the holes or depressions on the surface of the mucous membrane, are smaller apertures leading into recesses or follicles; these recesses are lined by mucous membrane, and are set round with closed capsules filled with a grayish fluid, and containing cells, and bodies like free nuclei. The capsules do not appear to have any apertures.

MUSCLES OF THE SOFT PALATE.

The muscles of the soft palate act as elevators and depressors. They are four in number on each side; and along the centre lies a thin fleshy slip, which is connected with the uvula.

G. Levator palati muscle.
H. Tensor palati muscle.
J. Azygos uvulæ muscle.

K. Superficial part of the palato-pharyngeus.
L. Deep part of the palato-pharyngeus.

The *elevator muscles*, two in number on each side, G and H, descend from the base of the skull, and enter the soft palate at their lower ends.

The *levator palati*, G, arises from the under surface of the apex of the temporal bone, and from the hinder part of the cartilage of the Eustachian tube; entering the pharynx above the upper constrictor (Plate xxiv.) it spreads out in the soft palate, forming a fleshy layer from the attached to the free edge, and unites with its fellow along the middle line.

The muscle contracting carries backwards and upwards the soft palate, placing this in a more horizontal position, and approaching the free edge and the uvula to the back of the pharynx. By that movement the part of the pharynx leading to the nose is much diminished; and if the upper constrictor muscle contracts at the same time the passage may be closed.

The *tensor vel circumflexus palati*, H, has a thin but wide origin from the skull, and from the fore part of the cartilage of the Eustachian tube—the cranial attachment reaching from the navicular fossa at the root of the internal pterygoid plate to the styloid process. Descending along the inner pterygoid plate, the muscle enters the pharynx between two points of attachment of the buccinator muscle (Plate xxiv.) and becoming tendinous, turns round the hamular process to be inserted partly into the os palati and partly into the aponeurosis of the palate beneath the muscles L and G. A small bursa exists where the tendon plays round the curved process of bone.

As this muscle is attached to the immovable hard palate its action must be more limited than that of the levator; it may assist the special elevator in bringing the side of the soft palate into a more horizontal position, and it will then fix and render tense the same part of the palate.

The two muscles above described are connected with the cartilaginous part of the Eustachian tube, and may act on it. Taking their fixed point below, they are enabled to open that tube which is ordinarily closed, and so to permit air to enter the cavity of the tympanum. During swallowing, and during forced expiration with the mouth and nose apertures closed, they act in the manner indicated; but some persons have the power of opening at will the Eustachian tube, and driving air in expiration into the tympanic cavity, without the nostrils being stopped.

Azygos uvulæ, J.—This slender muscle shortens the uvula and the middle part of the soft palate, and assists therefore the elevators. It consists of two slips of pale muscular fibres (only the right is seen), which arise above from the palate spine and the aponeurosis of the soft palate, and are inserted below into the submucous tissue of the uvula.

The *depressors* of the soft palate, two in number on each side, are directed downwards in the folds of the arches of the palate to the tongue and the thyroid cartilage.

The *palato-glossus* (constrictor isthmi faucium) lies in the anterior pillar, P. It is a thin narrow slip, which begins on the front of the soft palate, where it joins its fellow in the middle line; and ends on the side and dorsum of the tongue, as is shown in Plate xxvii.

If the lower end is fixed it can draw down the soft palate, stretching the same, so as to diminish the space between the tongue and the palate; and if both ends are fixed the muscle will be moved inwards towards its fellow, narrowing the isthmus of the fauces, as when a morsel of food is about to be swallowed.

The *palato-pharyngeus* is larger than the preceding and consists of two layers in the palate, which are separated by the levator palati and azygos uvulæ muscles.

The superficial thin layer, K, is close beneath the mucous membrane, and joins at the middle line the muscle of the other side. The deeper and stronger layer, L, unites with its fellow internally, whilst some of the upper fibres are fixed to the aponeurosis of the palate. Both layers meet at the outer border of the palate, and descend behind the tonsil in the fold, R, to be inserted mostly in the back of the thyroid cartilage, but a part blends in the pharynx with the upper constrictor.

Acting from below the muscle will bring down the arch, R, and will approach the same to the uvula; it will also draw down and back the soft palate towards the pharynx.

The *soft palate* from its position and its power of moving plays an important part in breathing, in the use of the blow-pipe, in swallowing, and in vomiting.

In breathing with the mouth open the air may pass through both mouth and nose, or only through the nose, according to the position of the movable palate. When the air obtains ingress and egress through both cavities at the same time the velum hangs vertically, as in the Drawing, and leaves a space between it and the tongue. When the air is transmitted only through the nose, the palate is applied to the back of the tongue, and shuts off the channel of the mouth.

During the use of a blow-pipe the mouth is first filled with air, and the soft palate is then applied to the back of the tongue to close the mouth behind, whilst the cheek-muscles force out from the oral space through the lips a continuous current of air. At intervals, however, the palate is raised temporarily during expiration for the purpose of refilling with air the cavity of the mouth.

In deglutition the soft palate directs the aliment into its downward channel. As soon as the morsel to be swallowed has reached the back of the tongue the movable palate is raised, and is arched over it so as to prevent its making an upward direction towards the nose. The depressor muscles contracting at the same time keep the flap fixed, and prevent its retroversion; and as the palato-glossus muscle moves inwards behind the morsel, barring with the tongue its return to the mouth, whilst the palato-pharyngeus forms with the uvula an inclined plane above it, the food is conveyed into the pharynx.

In vomiting the aliment takes a retrograde course from the stomach through the mouth; and the movable palate is used as a valve to shut off the upper region of the pharynx and the nose. The position of the velum during this act is similar to that occupied by it in deglutition, viz., it is moved somewhat horizontally backwards towards the wall of the pharynx, and the palato-pharyngei with the contracted uvula between them form behind an inclined plane. The soft palate is not capable, however, of blocking up entirely the tube of the pharynx, for some of the ejected ma ter is forced by the side of it into the nose cavity.

The influence of the soft palate on the voice seems to be small, though

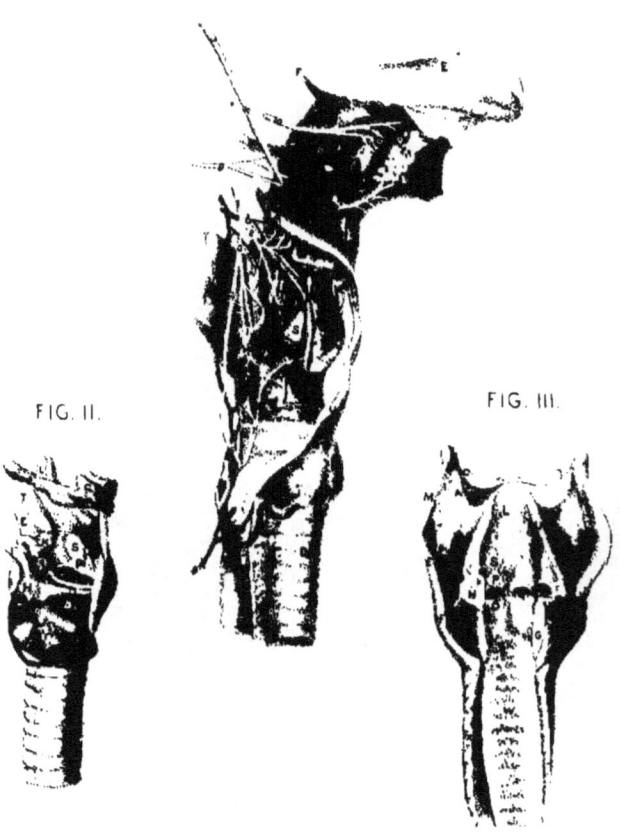

this flap forms part of the winding passage through which the sound is transmitted after its production by the vocal cords in the larynx. In the high notes in singing the palatine arches and the uvula are contracted, but touching them does not produce alteration of the note: this tense state has been thought to increase the resonance of the voice.

VESSELS AND NERVES.

The vessels and nerves appearing in this dissection have been noticed in the description of the preceding Plates.

- *a.* Ending of the external carotid artery.
- *b.* Temporal artery.
- *c.* Internal maxillary artery.
- *d.* Internal carotid artery.

- *e.* Inferior laryngeal branch.
- 1. Gustatory nerve.
- 2. Recurrent laryngeal nerve.

DESCRIPTION OF PLATE XXVII.

FIGURES ii. and iii. show the cartilages and ligaments of the larynx, with the vocal apparatus; and in Figure i. the muscles, vessels, and nerves are displayed.

In the preparation made for Figure ii., the muscles were removed, and the right half of the thyroid cartilage was cut off, except the fore part and the lower cornu; and then the muscles and the mucous membrane beneath the cartilage were taken away to lay bare the vocal cord, and the arytenoid cartilage of the same side.

Figure iii. exhibits the interior of the air passage in a larynx and windpipe slit down behind.

HYOID BONE AND THE CARTILAGES OF THE LARYNX.

The cartilages of the larynx can be studied with the aid of Figures ii. and iii.; and like parts in both Drawings are marked by the same letters of reference.

A. Great cornu of the hyoid bone.
B. Body of the os hyoides.
C. Small cornu of the hyoid.
D. Thyroid cartilage.
E. Upper cornu of the thyroid.
F. Lower cornu of the thyroid.
G. Cricoid cartilage.
H. Arytænoid cartilage.
I. Cartilage of Santorini.
J. Crico-arytænoideus posticus muscle.
K. Cuneiform cartilage.
L. Epiglottis.
M. Thyro-hyoid ligament.
N. Crico-thyroid ligament.
O. True chorda vocalis.
P. False chorda vocalis.
Q. Ventricle of the larynx.
R. Rima glottidis.
S. Sacculus laryngis.
T. Thyro-hyoid membrane.
U. Arytæno-epiglottid. fold.
V. Arytænoideus posticus muscl
W. Interior of the trachea.
X. Muscular part of the trachea.
Y. Rings of the trachea.

The *hyoid* or U-shaped *bone* is placed between the tongue and the larynx, to both of which it gives attachment. It consists of a central part or body, and of two lateral pieces on each side—the cornua.

The body, B, is the deepest part of the bone: it is convex and uneven in front, and concave and smooth behind. Elevator and depressor muscles are fixed into the fore part; and by its upper edge it gives attachment to the fibrous membrane of the tongue, and that of the larynx.

The cornua articulate with each side of the body. The large one, A, projects backwards behind the tongue, and is joined by muscles of the pharynx, larynx, and tongue. The small cornu, C, is a short rounded process, to which the stylo-hyoid ligament (X, Fig. i.) is connected.

Cartilages of the larynx. There are several pieces of cartilage in the larynx as in the trachea; but they differ in their nature. One set resembles the permanent cartilages of the ribs, and like them is prone to ossify; the other set, consisting of small pieces, is constructed of yellow cartilage, as in the eyelid, and is not transmuted into bone.

The large and firm cartilages, which are more or less ossified in the adult, are more immediately connected with the vocal cords: they are four in number, viz., the thyroid, cricoid, and two arytænoid.

The *thyroid* cartilage, D, is the largest and highest, and is named from protecting the rest like a shield. It is formed of two similar halves, which are widely separated behind, and are united in front at an acute angle, so as to be prominent beneath the skin (pomum Adami).

Each half ends posteriorly in a rounded thickened border, which is prolonged above and below into a point—the cornua: of these, the upper cornu, E, is the longest, and the lower one, F, articulates with the cricoid cartilage.

Externally muscles of the pharynx and larynx are fixed into the thyroid; and internally it receives the insertion of the vocal cords and of the muscles acting on those cords.

The *cricoid* cartilage, G, forms a ring around the air passage, and is much deeper behind than before, like a signet ring. On its upper border at the back are seated the two arytænoid cartilages; and outside and below these the lower cornua of the thyroid cartilage rest on it. Internally it is smooth and is lined by mucous membrane; and externally muscles are attached to it.

The part of the larynx inclosed by this cartilage is quite inextensible; and by means of the great depth of the cricoid behind, the arytænoid cartilages are raised to the height needful for the attachment of the vocal cords to them.

The *arytænoid* cartilages are something like a pitcher in shape, and are placed at the back of the larynx. Each is pyramidal in form, with the base resting on the upper border of the cricoid cartilage, and the apex blending with the cartilage of Santorini, I. Narrow and smooth internally or towards its fellow, it is widened and rough externally where muscles are inserted into it, Fig. ii., H. Its posterior part is hollowed, and lodges the arytænoid muscle; and from its fore part projects a spur into which the vocal cord, O, is fixed.

This is the most movable of the laryngeal cartilages; and as the vocal cord and most of the muscles altering the condition of that cord are connected with it, the production and modification of the voice are influenced by its position.

The remaining small cartilages do not take part in the production of the voice, though they may assist in modifying the same after it is formed, and they are therefore of secondary import. Five in number, the chief of them acts as a valve to the upper opening of the larynx, and is called epiglottis: the others are two pairs, one being named cartilages of Santorini, and the second, cartilages of Wrisberg.

The *cartilages of Santorini*, I (cornicula laryngis), are placed on the tops of the arytænoid cartilages. Wide below they gradually taper above, the points bending towards each other.

They bound posteriorly the upper laryngeal opening; and, enveloped by the mucous membrane, serve for the attachment of the folds, U, bounding laterally that opening.

The *cartilages of Wrisberg*, K, Fig. ii. (cuneiform cartilages), are placed in front of the cornicula, one in each arytæno-epiglottid. fold U. Each resembles a grain of rice in shape and size. The use of these is not known: they are not connected to the other cartilages by ligamentous bands.

The *epiglottis*, L, stands in front of the opening into the larynx (Plate xxvi.). Shaped like a leaf, with the wide part up and the pedicle down, it is attached by fibrous tissue to the thyroid cartilage. Its hinder or laryngeal surface has a smooth covering of mucous membrane with apertures for glands in its substance; and the fore part is connected to the tongue by a central and two lateral folds of mucous membrane. From each side is continued the arytæno-epiglottid. fold.

This valve is employed in closing the laryngeal opening during deglutition (p. 196). And when placed over the opening during the production of vocal sounds it causes the pitch of the note to be lowered.

ARTICULATIONS OF THE CARTILAGES.

The larger laryngeal cartilages are articulated together by means of joints where the extent of movement is great; and the larynx is further united to the hyoid bone above and the trachea below by fibrous membrane.

The *cricoid* and *thyroid cartilages* are articulated at two points, viz., laterally and in front.

Laterally there is a joint on each side between the lower cornu, F, of the thyroid and the side of the cricoid, in which an inclosing capsule and a lining synovial membrane are present. By means of this joint the front of the thyroid cartilage can be approximated to or removed from the cricoid. When the thyroid is depressed the vocal cords are tightened, and when it is raised or carried backwards they are relaxed.

Anteriorly a strong elastic membrane, *crico-thyroid*, N, closes the interval between the two. By its lower edge it is inserted into the upper border of the cricoid as far back as the arytænoid cartilage. Above it joins the lower border of the thyroid for a short distance, also the spur on the front of the arytænoid; and between those fixed points it forms a free edge, O, the vocal cord. This free upper edge can be tightened or rendered lax by the thyroid cartilage being depressed or raised.

Between the *arytænoid* and *cricoid cartilages* there is a very movable joint with a fibrous capsule surrounding the articular surfaces, and a synovial membrane lubricating them. In this joint the arytænoid cartilage can slide on the cricoid forwards and backwards, and inwards and outwards; and further, when the arytænoid is controlled by the muscles tending to draw it in opposite directions, it can be rotated round a vertical axis so as to move the anterior spur outwards and inwards.

The condition of the vocal cord, O, is altered by the movements of the cartilage. Thus it is relaxed when the arytænoid is carried forwards, and is tightened when the same is moved backwards; and the distance of the cords from one another will be increased and diminished as the two cartilages are moved from and towards each other. In rotation out the cords are separated and made tense, and in rotation in they are approached, but without being relaxed.

The smaller or accessory cartilages are articulated by ligamentous bands, but have not movable joints as in the larger cartilages.

The *cartilages of Santorini* are united to the top of the arytænoid by surrounding fibrous tissue: but at times there is some indication of a joint between the base of the one and the apex of the other.

The *epiglottis* is fixed below by a band (thyro-epiglottid) to the thyroid cartilage, close below the notch in the upper border; and in front it is united to the back of the hyoid bone by fibrous tissue—the hyo-epiglottid ligament.

The larynx joins the trachea below by a membrane similar to that connecting the rings of this tube; and it is attached to the hyoid bone above by the following ligament.

The *thyro-hyoid* membrane, T, is thin for the most part, but it forms rounded thicker cords behind—the thyro-hyoid ligaments, M. It is inserted below into the upper edge of the thyroid cartilage; but it is continued onwards to the upper edge of the os hyoides, muscles shutting it out from the lower edge of that bone: a synovial membrane intervenes between the two.

INTERIOR OF THE LARYNX AND THE VOCAL APPARATUS.

The larynx or the dilated upper part of the windpipe is wider above than below; and the space inclosed within the cartilages varies in form and size at different points. As a whole the larynx measures about one inch and a half from above down, one inch and a quarter across at the top, and about an inch across at the lower part.

The laryngeal cavity (Fig. iii.) communicates above with the pharynx by the epiglottid aperture, and below with the trachea. By means of muscles and the mucous membrane the space inside the thyroid cartilage decreases in width from the epiglottis to the level of the vocal cords, O, where only a narrow fissure—the glottis remains; but just above the vocal cord is a dilatation on each side, Q, which is named the ventricle of the larynx. Beyond the vocal cords the space enlarges to the size of the cricoid cartilage, and becomes circular. Its shape is something like an hourglass, the glottis, R, corresponding with the narrowest part of that instrument.

Vocal apparatus. Under this general term may be included the vocal cords, with the chink or interval between them; and the ventricle of the larynx and its pouch.

The *vocal cords* are two whitish bands on each side, which shine through the mucous membrane, and lie above and below the ventricular space, Q. Both are stretched between the thyroid cartilage in front, and the arytænoid behind.

The *upper band*, P, or the false vocal cord, forms a curve with the convexity upwards. In front it is fixed to the thyroid cartilage slightly above the middle; and behind to the outer part of the arytænoid. It consists of a bundle of white fibrous tissue, which is covered by the mucous membrane.

The use of this band is unknown. The voice is not produced by it, for it is removed so far from the centre of the laryngeal space as to be out of reach of the direct current of air.

The *lower* or *true cord*, O, Fig. ii., is stronger than the other, and is horizontal in direction. It is inserted in front into the thyroid cartilage about the centre of its depth, and behind into the anterior spur at the base of the arytænoid cartilage. In the male it measures rather more than half an inch, and in the female rather less. This band forms the upper free edge of the crico-thyroid ligament (Fig. ii., N), and consists of a bundle of fine elastic tissue covered by thin mucous membrane.

It has two free surfaces, one internal which looks to its fellow, and one above where it bounds the ventricle; and the free edge between those two surfaces is the part that is made to vibrate by the outgoing current of air.

Sound or voice is produced by the expired air throwing into vibration the free edges of the lower two vocal cords. In breathing the vibrating edges are at a distance from each other, and divergent behind, and the air passes by them without sound. In order that voice should be produced those edges require to be approximated and put parallel to each other by muscles, and so to be brought into the state called the vocalizing position.

The pitch of the voice varies with the degree of tightness or laxness of the vocal cords. If the cords are loose a deep sound ensues, but if they are tight, a high tone is formed. Alterations in the degree of tension depend upon the action of controlling muscles.

The *glottis* (rima glottidis), R, is the narrow interval or chink between the true vocal cords. Its extent is greater than that of the cords, for it reaches across the larynx; and it is bounded on each side by the vocal cord and the arytænoid cartilage. It measures from before back nearly an inch, and across at the base when dilated about a third of an inch: both measurements refer to the larynx of the male. In the female the size is less by two or three lines. During inspiration the space is larger than in expiration.

Its form changes with the dilatation. In a state of rest the interval resembles a spear-head with the shaft placed backwards; when dilated it is triangular in form, the base of the interval being behind.

The *ventricle* of the *larynx*, Q, is the hollow between the false and true vocal cords of the same side; and it extends from the thyroid to the arytænoid cartilage. The bottom of the hollow is wider than the opening into the larynx; and at its upper and anterior part it communicates with the sacculus laryngis, S. Into this hollow the mucous membrane sinks, and after lining the space, enters the laryngeal pouch.

This space by its position isolates the true vocal cord from the wall of the larynx, and permits the free-vibration of that band.

The laryngeal pouch (sacculus laryngis), S, is a small conical bag of the mucous membrane, which projects upwards from the ventricle of the larynx, and when distended reaches as high as the upper border of the thyroid cartilage. Fig. ii. gives an inner view of its position on the side of the epiglottis; and in Fig. i. it is seen from the outside as it rises above the thyro-arytænoid muscle, P.

Closed and dilated above, the pouch is narrow below; and it opens into the ventricle by a small hole, which is diminished somewhat by a projection of the mucous membrane. Over the outer surface are scattered numerous mucous glands (sixty or seventy in number) which open by small ducts on the inner surface, and pour their secretion over the contiguous parts, viz., the ventricle and the vocal cords.

The *mucous lining* of the larynx forms a fold, U (arytæno-epiglottid), on each side of the upper orifice, and extends through the cavity to the trachea. Furnishing a very thin covering without glands to the vocal cords, it sinks into the ventricle between them, and gives rise to the sacculus. As low as the vocal cords it is loosely united to the subjacent parts by areolar tissue, but it is joined closely to those bands without the intervention of any submucous stratum. In consequence of the closeness of its attachment to the cords the swelling from fluid effused into the areolar tissue in œdema of the glottis does not extend below that point; and thus, though the upper orifice of the larynx may be closed by the swelling, air can be admitted to the lungs by an artifical aperture through the crico-thyroid membrane, N, as in the operation of laryngotomy, because this opening will be situate below the swollen parts.

FIGURE I.—For this Drawing the dissection was prepared by removing the greater part of the right half of the thyroid cartilage, and then taking the areolar tissue from the subjacent muscles, vessels, and nerves. Some nerves which enter the mucous membrane behind the larynx from both laryngeal trunks could not be preserved.

On the right side of the tongue the extrinsic muscles have been defined as they enter it.

In this, as in the other Figures, the hyoid bone, the cartilages of the larynx with some ligaments, and the trachea and the thyroid body are depicted.

A. Os hyoides.
B. Thyroid cartilage.
C. Cricoid cartilage.
D. Trachea.

E. The tongue.
F. Palato-glossus muscle.
G. Stylo-glossus.
H. Pharyngeo-glossus.

I. Cornicula laryngis.
J. Crico-thyroid membrane.
K. Hyo-glossus muscle.
L. The epiglottis.
M. Genio-hyo-glossus.
N. Thyro-hyoid membrane.

S. Sacculus laryngis.
T. Thyroid body.
U. Pyramid of the thyroid body.
W. Levator glandulæ thyreoideæ.
X. Stylo-hyoid ligament, ossified.
Z. Upper part of the œsophagus.

MUSCLES OF THE LARYNX.

Some of the intrinsic laryngeal muscles act more immediately on the arytænoid cartilages, approximating them to, or removing them from each other, and control the width of the glottis. Others make tense or lax the vocal cords, and so govern the pitch of the voice. One pair of muscles depresses the epiglottis.

O. Depressor of the epiglottis.
P. Thyro-arytænoideus.
Q. Crico-arytænoideus lateralis.

R. Crico-arytænoideus posticus.
V. Arytænoideus.
Y. Crico-thyroideus, cut.

Muscles governing the size of the glottis.—The interval between the vocal cords can be widened or narrowed by the three following muscles.

The *crico-arytænoideus posticus*, R (J, Fig. iii.), arises from the right lateral depression on the back of the cricoid cartilage, and is inserted above into the base of the arytænoid cartilage at the outer side.

When this muscle acts the arytænoid cartilage will be rotated around its vertical axis, and the anterior spur will be moved outwards away from the middle line. By this movement the glottis is widened at the base, and the upper aperture of the larynx is also made larger.

The *crico-arytænoideus lateralis*, Q, arises from the upper edge of the cricoid cartilage, at the lateral aspect; and taking a backward direction it is inserted with the preceding into the external prominence at the base of the arytænoid cartilage, and into the contiguous part of the outer surface.

As the preceding muscle moves outwards the external projection of the cartilage, the lateral crico-arytænoideus is put on the stretch; but as soon as the posterior muscle ceases to contract, the lateral one will restore the displaced cartilage to its usual position. This muscle, acting by itself, will turn inwards the anterior spur, and diminish the width of the glottis.

The *arytænoideus*, V, the only single muscle of the larynx, closes the interval between the arytænoid cartilages. It consists mostly of transverse fibres, which are attached to the hollowed posterior surfaces of the cartilages; but it possesses also two superficial bands, which are directed from the base of one cartilage to the apex of the other. These oblique slips cross each other at the middle, and join in front the thyro-arytænoideus and the depressor epiglottidis.

The fibres of the muscle contracting will draw the arytænoid cartilages towards each other, and diminish the width of the glottis. And, as this movement approximates the vocal cords, the muscle is one of the two employed in placing the cords in the vocalizing position. The muscle diminishes behind the width of the upper laryngeal orifice.

Muscles governing the pitch of the voice.—The muscles making tight or loose the vocal cords, and rendering the voice either high or deep in tone, are the two subjoined.

The *thyro-arytænoideus* muscle, P, lies outside the vocal cord of the same side, to which it is closely united. Anteriorly it arises from the lower half (in depth) of the thyroid cartilage, and from the contiguous crico-thyroid membrane; and it is inserted behind into the base and outer surface of the arytænoid cartilage. Its inner and lower fibres are transverse, but the outer ascend and join the depressor of the epiglottis, O.

Through the action of this muscle the arytænoid will be drawn forwards towards the thyroid cartilage, and the vocal cord of the same side will be relaxed, as when deep or grave sounds are produced. The muscle is supposed (Willis) to have the power of placing the inner vibrating edge of the vocal cord parallel to its fellow.

The *crico-thyroid* muscle, Y, can be seen entire in Plate xxiv. Placed on the front of the larynx, it arises from the side and fore part of the cricoid cartilage; and it is inserted into the inferior cornu, and the lower border of the thyroid cartilage nearly to the middle line.

Supposing the attachment to the cricoid cartilage to be the fixed point, the muscles of opposite sides will bring down the thyroid cartilage in front. By this movement the interval between the arytænoid and thyroid cartilages is increased, and consequently the vocal cords are tightened, and put into the state necessary for the production of a high note. If the thyroid is supposed the fixed point, the front of the cricoid will be raised, whilst the back of the same with the arytænoid cartilages will be lowered, and the vocal cords will be likewise stretched.

The *depressor of the epiglottis,* O (thyro-arytæno-epiglottideus), is a thin and indistinct layer of muscular fibres, which is contained in the arytæno-epiglottid fold, U, and consists usually of two parts. The chief bundle of fibres comes from the top of the arytænoid cartilage, where it is continuous with the thyro-arytænoideus and arytænoideus muscles: and the other slip is attached to the thyroid cartilage near the insertion of the epiglottis. The fibres of the muscle ascend on the side of the opening of the larynx, and are inserted into the margin of the epiglottis.

The lower fibres of the muscle cross the top of the sacculus laryngis, and are supposed by Mr. Hilton to compress the sac: this part has been named by him *arytæno-epiglottideus inferior.**

In swallowing the epiglottis may be lowered by the action of the muscles of both sides, after the larynx has been elevated; and the laryngeal orifice can be diminished by the shortening and moving inwards of the arytæno-epiglottid fold. In the production of very deep notes the muscles draw down the epiglottis over the aperture of the larynx.

NERVES OF THE LARYNX.

There are two laryngeal nerves on each side, the superior and inferior. One is supplied nearly altogether to the mucous membrane, and the other chiefly to muscles.

1. Upper laryngeal nerve.
2. Branches to the mucous membrane of the larynx.
3. Branch for the arytænoideus.
4. Branch to join inferior laryngeal.
5. Inferior laryngeal or recurrent nerve.
6. Branch to join upper laryngeal.
7. Branch to muscles.
8. Hypoglossal nerve.
9. Glosso-pharyngeal nerve.
10. Gustatory nerve.

The *upper laryngeal* nerve, 1, pierces the thyro-hyoid membrane, and divides into branches. From the branch, 2, offsets are distributed to the root of the tongue, and to the mucous membrane of the larynx; between the border of the epiglottis and the true vocal cord, one or two pierce the depressor of the epiglottis. The branch 3 enters the arytænoideus mus-

---

* Description of the sacculus or pouch in the human larynx. By Mr. John Hilton. Guy's Hospital Reports, vol. 2. Lond. 1837, p. 519.

cle, V, and supplying it, passes through to the mucous lining of the larynx. From the branch, 4, offsets are furnished to the pharyngeal mucous membrane; and this joins finally the recurrent laryngeal nerve.

Before the nerve enters the larynx it gives off high in the neck the external laryngeal branch (Plate xxiv. 4), which ends in the crico-thyroideus muscle, Y, supplying it entirely.

The upper laryngeal is the sensory nerve of the mucous membrane of the larynx as low as the true vocal cord; and by its extreme sensibility it guards the upper part of the passage against the entrance of anything but the air. As soon as a particle of food or drink touches the lining membrane, the respiratory muscles are called into play by a reflex act, and the foreign body is expelled by coughing. In the attempt to breathe an irrespirable gas the passage is closed by the contraction of the surrounding muscles, also through a reflex act. When the nerve is cut across in an animal during life the sensibility of the part is lost, and food may enter the larynx.

To the crico-thyroideus muscle, which it supplies alone, it gives motor influence as well as sensibility; and to the arytænoideus, to which with the recurrent it furnishes offsets, it imparts only sensibility.

The *inferior laryngeal* or recurrent nerve, 5, ascends over the side of the cricoid cartilage, and ends in muscular offsets beneath the thyroid. At first the nerve supplies branches to the mucous membrane of the pharynx, and the communicating branch, 6, which joins the upper laryngeal under the thyroid cartilage. The continuation of the nerve, 7, then terminates in branches for muscles:—one belongs to the crico-arytænoideus posticus, R; a second, which passes beneath the preceding muscle, enters the arytænoideus, V; and another gives nerves to the crico-arytænoideus lateralis, Q, and the thyro-arytænoideus, P. In short, the nerve supplies all the special laryngeal muscles except the crico-thyroideus, Y, which receives the external laryngeal branch of the superior laryngeal nerve.*

---

* Anatomists are silent for the most part respecting the nerve to the muscle here called depressor of the epiglottis; but Mr. Hilton states (Guy's Hospital Reports, vol. 2, 1837) as the result of "repeated and careful dissections" that it is supplied from the recurrent nerve by means of two filaments which are prolonged from the branch of the same nerve to the thyro-arytænoideus. Neither in my own dissections, nor in those of Mr. P. B. Mason and Mr. J. S. Cluff, formerly Demonstrators of Anatomy, could any separate branch be traced from the recur-

The recurrent is the motor nerve of the muscles acting on the vocal cords, to all of which, except to the crico-thyroideus, it gives branches. But it must bestow sensibility by means of the offsets ramifying in the mucous membrane.

If the recurrent nerves are cut through, the muscles are paralyzed; and as the vocal cords cannot be placed in the vocalizing position, and cannot receive the necessary degree of laxity or tension, voice will not be produced.

VESSELS OF THE LARYNX.

Two arteries on each side, which are companions to the nerves, ramify in the larynx; they are named upper and lower laryngeal. Other small arteries from the upper thyroid enter the larynx, below, by perforating the crico-thyroid membrane.

a. Upper laryngeal artery.
b. Ascending branch ⎫ of the upper
c. Descending branch ⎬ artery.
d. Communicating branch of the upper laryngeal.
e. Communicating branch of the lower laryngeal artery.
f. Muscular branch of lower laryngeal.
g. Inferior laryngeal artery.
h. Branches of superior thyroid artery to the thyroid body.
k. Branches of inferior thyroid artery to the under part of the thyroid body.

The *upper laryngeal* artery, a, resembles the nerve of the same name in its branches, but it is not distributed so exclusively to the mucous membrane. The offsets, b and c, supply the mucous membrane from the root of the tongue to the chorda vocalis; and from c, arteries are furnished to the muscles, O, P, and Q, under the thyroid cartilage, and to the crico-thyroideus, Y. The branch, d, anastomoses with the inferior laryngeal both under the thyroid cartilage, and in the mucous membrane of the pharynx.

The *inferior laryngeal*, g, gives branches to the posterior laryngeal muscles, R and V, and to Q in part; and it joins the upper laryngeal outside the thyro-arytænoideus muscle, P. Branches of it enter the mucous membrane of the pharynx, and communicate again with the upper laryngeal by the offset, e.

---

rent nerve to the muscle. Mr. Cluff made six special examinations of the human larynx, one of the larynx of a donkey, and one of the larynx of a cat.

Veins accompany the arteries. The upper laryngeal opens through the superior thyroid vein into the internal jugular trunk; and the lower sends its blood into the innominate vein along the inferior thyroid branch.

THE THYROID BODY AND THE TRACHEA.

In a side view, the thyroid body, T, is only partly visible. This organ is larger in the female than in the male, and is more developed in the fœtus than in the adult relatively to the rest of the body: its use is not known.

It is placed opposite the upper part of the trachea; and consists of two lobes, right and left, which are firmly attached to the windpipe, and project upwards, one on each side, as far as the thyroid cartilage. A narrow part, the isthmus, joins the lobes below in front of the trachea. Each lobe is pointed above and wide below; and it lies between the larynx and the common carotid artery, where it is covered by the depressor muscles of the hyoid bone (Plate xxiv.).

Projecting upwards from the left lobe, or from the isthmus, is a small tapering part, U—the pyramid, which is connected to the os hyoides by a band of fibrous tissue. Sometimes, as in the Drawing, a thin muscular slip, W, *levator glandulæ thyroideæ*, unites the pyramid with the hyoid bone.

Brownish red or purplish in color, it consists of small masses or lobules about as large as the little finger nail. It does not possess any excretory duct. On cutting into it a thick yellowish fluid escapes from small closed capsules or vesicles.

The swelling of the throat known as a wen or Derbyshire neck, is caused by enlargement of the thyroid body.

Bloodvessels.—Two large arteries on each side ramify in this body. The *upper thyroid*, h, a branch of the external carotid, enters the apex of the lobe, but it distributes some branches over the surface, which join the other arteries. The *lower thyroid*, k, is usually larger than the upper, and is a branch of the subclavian trunk: it penetrates the base of the lobe, and offsets ramify over the under surface. All the arteries communicate freely together.

Three large *thyroid veins* issue on each side. Two, upper and lower thyroid, run with the arteries of the same name, and end—the former in the internal jugular, and the latter in the innominate vein. A middle

FIG. I.

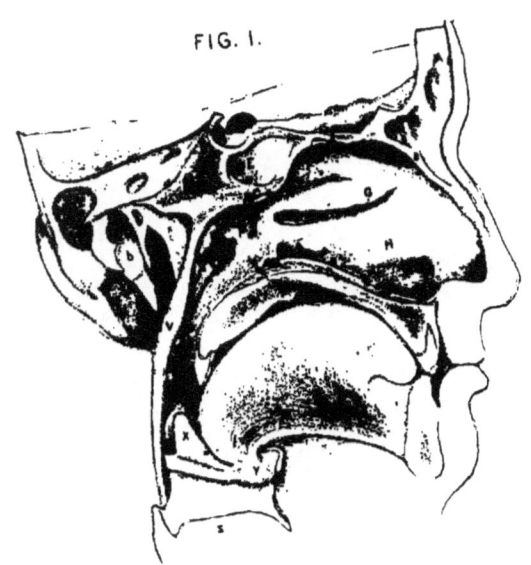

FIG. II.

FIG. III.

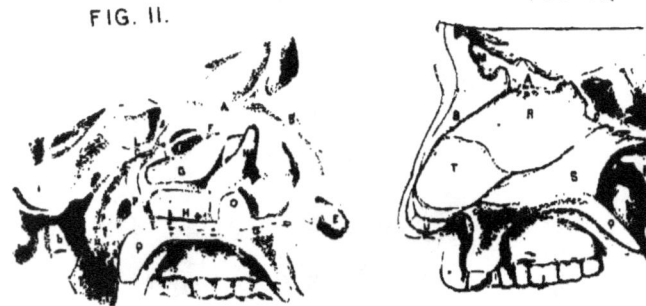

thyroid vein leaves the middle of the lobe, and crossing the common carotid artery joins the internal jugular trunk.

The *trachea* or windpipe, D, reaches from the larynx to the thorax, and divides in that cavity into two pieces or bronchi—one for each lung. Placed in front of the œsophagus, Z, along the middle line of the body, it is round and firm in front, but flat and soft behind, and is always pervious to the air. Its transverse width is about an inch in the male, but less in the female. Its fore and hinder parts differ much in their composition.

The firm fore part of the tube consists of dense fibrous membrane, which incases separate pieces of cartilage about one-sixth of an inch wide, and forming three-fourths of a circle. Each piece has its convexity directed forwards; and the whole keep apart the walls of the tube. Behind, where the trachea is flattened, it is constructed by fibrous membrane (X, Fig. iii.) continuous with that containing the pieces of cartilage; and beneath it is a layer of transverse muscular fibres, together with some superficial bundles of short longitudinal fibres.

Lining the trachea is a mucous membrane covered with a columnar and ciliated epithelium; and beneath the same is a layer of elastic tissue which is collected into bundles in the flat part of the tube. Many glands are placed beneath the mucous membrane; and the largest occupy the back of the windpipe, where some are external to the fibrous and the muscular layer.

DESCRIPTION OF PLATE XXVIII.

These three Figures of vertical sections of the nose will indicate the boundaries of that cavity, and the openings into it.

For Figure i., the right half of the nasal cavity was cut through vertically, and the septum nasi was removed—the fore part of the skull having been previously detached for the dissection of the pharynx.

The nose was sawed through on the left of the septum for Figure ii.; and pieces of the middle and inferior spongy bones were cut out to render evident the openings into the meatuses.

And for Figure iii. the mucous membrane was removed from the septum nasi, after the saw had been carried vertically through the left nasal fossa.

BOUNDARIES OF THE NASAL CAVITY.

Some of the boundaries appear in all the Figures, and the same letters of reference are used for them.

A. Middle part of the roof of the nasal fossa.
B. Fore part of the roof.
C. Back of the roof.
D. Floor of the nasal cavity.
E. Dilatation within the nostril.
F. Upper spongy bone.
G. Middle spongy cone.
H. Lower spongy bone.
I. Upper meatus of the nose.
J. Middle meatus.
K. Lower meatus.
L. Sphenoidal sinus.
M. Frontal sinus.

N. Funnel-shaped prominence of the ethmoid bone.
O. Aperture of the nasal duct.
P. Opening of the Eustachian tube.
Q. Soft palate cut through.
R. Descending plate of the ethmoid.
S. Vomer.
T. Cartilage of the septum.
U. Cartilage of the aperture.
V. Pharynx.
W. Genio-hyo-glossus muscle.
X. Epiglottis.
Y. Os hyoides.
Z. Thyroid cartilage.

Each half of the nose cavity is a flattened space, which communicates with the face and the pharynx, and with the hollows or sinuses in the surrounding bones. It intervenes between the base of the skull and the mouth, and occupies the interval between the eye sockets. The bones of the face and skull enter into its construction; and the boundaries are named roof and floor, inner and outer wall.

The floor, D, is horizontal and smooth; and its bony framework consists of the palate pieces of the upper maxillary and palate bones.

The roof reaches from the nostril to the posterior naris, and is sloped before and behind. Its centre is formed by the thin cribriform plate of the ethmoid, A, and is nearly straight; the fore part is made up of the frontal and nasal bones, and the lateral cartilage; and the hinder part is bounded by the anterior and inferior surfaces of the body of the sphenoid, with the sphenoidal spongy and the palate bones.

The inner wall is the septum or partition between the fossæ of opposite sides. In it are the descending plate of the ethmoid, R, the vomer, S, and the triangular cartilage, T.

The outer wall is marked by projecting osseous pieces with subjacent hollows, and is constructed of several bones. From before back the following is the order of succession, viz.: os nasi, upper jaw, lachrymal, ethmoid, and palate bones, with the internal pterygoid plate of the sphenoid bone. Below the nasal, in front, the lateral cartilage is found.

The width of the nasal fossa is larger below than above; and at the floor close to the septum is the greatest space available for passing an instrument through the cavity. Across the upper part of the fossa the spongy bones project, so as nearly to touch the septum. From before back the length measures about three inches along the floor, and the depth amounts to two inches at the centre.

In front is the opening called the nostril: this is an elongated hole which is surrounded, except behind, by the cartilage of the aperture, U, and is always open. For the distance of half an inch within the nostril is a dilatation, E, large enough to take the end of the finger, which is lined by skin provided with hairs or vibrissæ. Behind, the space communicates with the pharynx by the posterior naris (Plate xxvi.).

In breathing the air passes ordinarily through the lower half of the nasal fossa, but by sniffing, as in the attempt to recognize faint odors, the current can be directed upwards to the region where the olfactory nerve ramifies.

Through the lower part of the nasal fossa the opening of the Eustachian tube, P, can be reached. To enter that tube an instrument should have the requisite size and curve, and should be directed along the floor close to the septum until it reaches the posterior naris; then the point is to be turned upwards and outwards into the aperture. In like manner a flexible tube can be passed through the cavity to the pharynx for the purpose of conveying liquid food into the stomach.

Blood escaping into the nasal cavity from rupture of the vessels of the mucous lining requires to be confined within the space when the loss of a fresh quantity may be injurious to health or endanger life. In closing the nasal fossa the posterior naris is stopped first by a plug inserted through the mouth in the following manner:—An elongated dossil of lint or cotton wool of the size of the opening is to have a piece of silk or small twine tied around the middle, as as to leave the ends about a foot long. Next, a bit of wire (not too stiff), with a noose at the end and rather curved downwards, is to be pushed along the floor of the nose and behind the soft palate until it can be seen through the open mouth. One of the string ends should be inserted through the noose with a pair of forceps, the other being retained in the left hand. By withdrawing the wire the string will be brought out through the nostril; and by means of that piece of string the plug can be dragged through the mouth, and round the soft palate with the aid of the left forefinger to the posterior naris. The two strings may be then tied between the nose and mouth. Finally to complete the closure of the nasal cavity the nostril is to have a plug inserted into it.

When the surgeon considers the bleeding not likely to return the plugs are to be taken away. For the removal of the anterior one the proceeding is simple; but the posterior has to be taken out through the mouth in this way:—The knot on the face being untied, the plug is to be dislodged from the posterior naris by pulling downwards and backwards with a forceps the string in the mouth; and it is then to be conducted round the soft palate to the exterior of the body.

SPONGY BONES AND THE MEATUSES.

Three curved bones, Fig. i., project into the nasal cavity from the outer wall; they are named from their form spongy or turbinate; and from their position, upper, middle, and lower. These osseous pieces do not extend the whole length of the outer wall, but are confined to a part limited by two lines continued upwards—one from the front and the other from the back of the hard palate.

The *upper spongy* bone, F, is a process of the lateral mass of the os ethmoides; and it occupies the posterior half of the interval before mentioned.

The *middle spongy* bone, G, is also a process of the ethmoid, and forms the lower curved edge of the lateral mass of that bone: usually it reaches all across the space included by the two vertical lines.

The *inferior spongy* bone, H, one of the facial bones, is larger than the others, and its length rather exceeds the limits of the space referred to.

The spongy bones are thin and brittle; and as they are convex on the inner surface and concave on the outer, channels or meatuses exist between them and the wall to which they are attached. They are covered by the mucous membrane, and afford greater surface for the ramifications of the nerves and bloodvessels.

The *meatuses* Fig. i., are the lengthened spaces between the spongy bones and the outer wall; and they are the same in number as those bones. Occasionally there is a rudiment of a fourth space above the rest, as in Fig. ii.

The upper meatus, I, less deep and long than the others, communicates with the posterior ethmoidal cells by an aperture or apertures at the fore part (Fig. ii.).

The middle meatus, J, has several openings in it from hollows in the surrounding bones; and in Fig. ii. the middle spongy bone is represented cut through to show the apertures. At the front of the meatus is an elongated eminence, N, of the ethmoid bone, with two grooves, one before and the other behind it: the anterior groove leads upwards into the frontal sinus, M, and the posterior opens into the anterior ethmoidal cells. Close above the lower part of the prominence referred to, and midway between the letters J and N, is the small round hole of the antrum maxillare.

The inferior meatus, K, receives the ductus ad nasum; and to see this the lower spongy bone will require to be cut through in front. In the dried bone the canal for the tears has a wide funnel-shaped end in the meatus; but in the recent state a piece of the lining membrane of the nose is stretched over the aperture forming a valve for it, and leaves only a small oblique passage for the tears. In the Figure the size of the opening, O, is to be seen. Usually the flap closes the aperture, and prevents air from being driven out of the nose into the lachrymal canals. An instrument entering the duct from below must necessarily injure the valve.

One sinus, viz., that of the body of the sphenoid bone, L, does not open into a meatus; its aperture, which is rather large, may be seen on the slanting hinder part of the roof.

THE MUCOUS MEMBRANE AND THE BLOODVESSELS.

The mucous membrane, named pituitary and Schneiderian, clothes the cavity, uniting with the periosteum of the bones, and joins the skin in front, and the lining of the pharynx posteriorly. It is continued over the foramina transmitting vessels and nerves into the cavity, so as to close them; but it sinks into the apertures leading into the sinuses in the surrounding bones, and lines those air spaces, whilst it diminishes somewhat the size of their openings. Through the nasal duct it is continued upwards to the lachrymal sac, and forms below a thin valve, O, which shuts the opening.

In the lower half of the nasal cavity the membrane is thick and vascular, particularly over the septum nasi and the lower spongy bone: and it increases the surface of the latter by being prolonged from the lower margin. Its epithelium is columnar and ciliated, except in the dilatation near the nostril where it joins the epidermis and is laminar: at this same spot it is provided with papillæ, and with long hairs or vibrissæ. Large mucous glands abound in the lower part of the nose, and their apertures cover the surface.

In the upper part of the nose the mucous layer is less thick and vascular, and is of a yellowish color. The epithelium is thick, especially over the olfactory region, and is laminar according to Bowman; though other observers state that it is ciliated at spots, and is columnar. The glands are numerous. In the olfactory region these resemble the sweat-glands of the skin, and open in rows between the nerve branches: their long ducts are lined by scaly epithelium.

Bloodvessels.—As the *arteries* are not injected, suffice it to say that they are derived chiefly from the internal maxillary, and come through the spheno-palatine foramen. A few enter through the apertures in the roof from the ophthalmic artery, and near the nostril are branches of the facial. In the pituitary membrane, they form a network, and on the surface and free edges of the two lower spongy bones they ramify in plexuses beneath the membrane.

The *veins* accompany the arteries, and form large venous plexuses on the septum nasi and the middle and lower spongy bones. Through the apertures in the cribriform plate of the ethmoid, the veins of the nasal cavity communicate with those in the cranium.

THE OLFACTORY REGION AND THE NASAL NERVES.

In the mucous membrane at the top of the nasal cavity the olfactory nerve ends, and the seat of smelling is located. To this part the term olfactory region has been applied by Mr. Bowman. Its situation is under the cribriform plate of the ethmoid bone, and it extends down for about an inch on the septum nasi and the outer wall.

Over the limits of this region the mucous membrane is thin, as before said, with thick scaly epithelium, and the glands are like sweat-glands.

The vessels construct a network in the adult, but in the fœtus, Mr. Bowman found on injecting them loops here and there with enlargements, suggesting to him the idea of rudimentary papillæ.

Olfactory nerve.—The offsets of the olfactory nerve enter the nose through the foramina in the cribriform plate of the ethmoid bone, and penetrating the mucous membrane, they divide and subdivide in a plexiform manner till they are reduced to the necessary degree of fineness, but the mode of ending of the nerve-filaments is not known. Recent researches (Schultze) point to the ending of the branches in olfactory or nerve-cells, which resemble somewhat columnar epithelium, and project to the free surface amongst the cells of the epithelium.* In their structure the nerve fibrils resemble the sympathetic more than other nerves, for they are granular, and are provided with oval corpuscles, which become visible on the addition of acetic acid.

Upon this nerve the faculty of recognizing odors depends. In ordinary breathing, when the air traverses chiefly the lower half of the nasal cavity, faint odors fail to give indication of their presence; but if the air is carried upwards into the olfactory region by sniffing, the odorous particles diffused in the air will be detected, because they are brought more completely into contact with the nerves. Touching the olfactory region with a solid body, as with a probe, does not excite the sensation of smell. Disease of the brain sometimes gives origin to supposed offensive odors.

Fifth Nerve.—Through the following offsets of the first and second trunks of the fifth nerve the pituitary membrane is supplied.

The nasal nerve of the ophthalmic trunk ramifies in the fore part of the cavity from the roof to the nostril, and acts as the guardian nerve of the anterior opening by endowing the part referred to with great sensibility. Irritation of the anterior portion of the nasal cavity gives rise through this nerve to the reflex act of sneezing, with the view of dislodging the unusual stimulus by a strong current of air rapidly expelled.

The spheno-palatine branches of the upper maxillary nerve furnish offsets through Meckel's ganglion to all the remainder of the cavity; these branches pass for the most part through the spheno-palatine foramen. Common sensibility and the nutrition of the mucous membrane are dependent upon this trunk of the fifth nerve.

---

* Manual of Human Microscopic Anatomy. By A. Kölliker. Lond., 1860, p. 604. In this work reference is given to the writings of Herr Schultze.

INDEX.

ABDUCENS nerve, 108, 120
Abductor indicis, 88
    minimi digiti manus, 87
    pollicis manus, 86
Acromial thoracic artery, 23
Adductor minimi digiti. 87
    pollicis, 86
Anastomosis of arteries
  in the axilla, 23
  at the elbow, 46
  in the hand, 81, 82, 89
Anastomotic artery of brachial, 45
Anconeus muscle, 94
Annular ligament of wrist, posterior, 95
Aperture of Eustachian tube, 194, 217
    larynx, 196
    nares, 195
    œsophagus, 196
Aponeurosis of soft palate, 197
Arch, palmar, deep, 89
    superficial, 81
  palatine, 197
  of subclavian, 154
Arm, dissection of, 37
Artery, anastomotic brachial, 45
  auricular posterior, 150
  axillary, 5, 19
  brachial, 34, 41
  buccal, 166
  carotid, common, 146
    external, 149
    internal, 109, 184
  carpal, ulnar, anterior, 76
    posterior, 97
    radial, anterior, 69
    posterior, 97
  central of the retina, 113
  cervical ascending, 156
    deep, 161
    occipital, 161
  ciliary anterior, 113
    posterior, 113

Artery, circumflex, anterior, 8, 52
    posterior, 8, 52, 61
  companion of median nerve, 76
  crico-thyroid, 183
  dental anterior, 178
    inferior, 165
    posterior, 166, 178
  digital of hand, 82, 89
  dorsal of index finger, 97
    of scapula, 23, 53
    of tongue, 175
    of thumb, 97
    of wrist radial, 69
    ulnar, 97
  ethmoidal, 113
  facial, 149, 166, 183
  frontal, 113
  infra orbital, 178
  intercostal upper, 156
  interosseous anterior, 76, 101
    posterior, 100
    of hand, 89, 97
  labial inferior, 166
  lachrymal, 113
  large of thumb, 89
  laryngeal inferior, 192
    superior, 192
  lingual, 150, 175
  mammary internal, 156
    external, 8
  masseteric, 166
  maxillary internal, 149, 165
  median, 76
  meningeal, large, 110, 166, 169
    small, 169
  metacarpal ulnar, 76
    radial, 97
  mylo-hyoid, 166
  nasal, 113
  nutritious of humerus, 45
  occipital, 149, 161
  ophthalmic, 113

Artery, palatine inferior, 183
    palmar deep, 89
    palpebral, 113
    perforating, of hand, 89
    pharyngeal ascending, 183
    profunda cervical, 156, 161
        humeral, 45, 60
    radial, 68, 88
    ranine, 175
    recurrent interosseous, 101
        radial, 69, 101
        ulnar, 76
    scapular posterior, 130
    subclavian, 127, 154
    sublingual, 175
    submental, 150
    subscapular, 8, 23
    superficial of palm, 69
    supra orbital, 113
        scapular, 130, 156
    temporal, 149
        deep, 166
    thoracic acromial, 23
        alar, 8, 23
        long, 8, 23
        superior, 23
    thyroid inferior, 156, 182
        superior, 149, 183
    tonsillitic, 183
    transverse cervical, 131, 156
        facial, 149, 178
    tympanic, 184
    ulnar, 69, 74, 81
    vertebral, 109, 156, 161
Articulation of laryngeal cartilages, 204
Arytænoid cartilages, 203
    muscle, 210
Ascending cervical vessels, 156
    pharyngeal vessels, 183
Auditory nerve, 108
Auricular artery, posterior, 149
    nerve, large, 133, 152
        posterior, 134
Auriculo-temporal nerve, 152, 170
Axilla, 3
    dissection of, 1
Axillary artery, 5, 19
    glands, 12
    plexus, 10
    sheath, 19
    vein, 9, 24
Axis, thyroid, 156

Azygos uvulæ muscle, 199

BASILIC vein, 29, 40
Biceps humeralis, 18
Blood-letting at elbow, 30
Brachial aponeurosis, 27, 38
    artery, 41
    plexus, 25, 189
    veins, 40
Brachialis anticus, 39
Buccal artery, 166
    nerve, 168
Buccinator muscle, 164

CARDIAC nerves, lower, 188
        middle, 188
        upper, 188
Carotid artery, common, 146
    external, 149
    internal, 109, 148, 184
Carpal arteries, radial, 69, 97
    ulnar, 76
Cartilage, arytænoid, 203
    cricoid 203
    cuneiform, 204
    thyroid, 202
    triangular nasal, 216
Cartilages of nose, 216
    of Santorini, 203
    of trachea, 215
    of Wrisberg, 204
Cavernous sinus, 116
Central artery of retina, 113
Cephalic vein, 24, 30, 40
Cervical ganglion, inferior, 188
    middle, 188
    superior, 187
    nerves, anterior, 161, 189
        posterior, 161
    plexus, 133, 189
        deep branches, 189
        superficial, 133
Cervico-facial nerve, 140, 151
Chiasma of optic nerves, 106
Chorda tympani nerve, 171
Chordæ vocales, 206
Ciliary arteries, 113
    nerves, nasal, 115
        lenticular, 119
Circular sinus, 116
Circumflex artery, anterior, 8, 52
    posterior, 8, 52, 61

Circumflex nerve, 11, 53
Comes nervi mediani, 77
Commissure of optic nerves, 106
Complexus muscle, 159
Compression of arteries
    brachial, 41
    femoral, 33
    subclavian, 128
Constrictor inferior, 191
    faucium, 199
    middle, 191
    superior, 191
Coraco-brachialis, 3, 18
    clavicular ligament, 14
Cords, vocal, 206
Cornicula laryngis, 203
Costo-coracoid membrane, 19
Cranial nerves, 106
Crico-arytænoid joint, 205
    muscle lateral, 209
      posterior, 209
  thyroid joint, 204
    membrane, 204
    muscle, 210
Cricoid cartilage, 203
Cuneiform cartilages, 204
    articulations of, 205
Cutaneous nerves of arm, 74
    face, 179
    hand, palm, 70, 77
    neck, front, 140
    shoulder, 53
    thorax, 11
Cutaneous veins of elbow, 28

DEEP cervical artery, 156, 161
  facial vein, 167
Deglutition, act of, 192, 196
Deltoid muscle, 51
Dental artery, anterior, 178
    inferior, 165
    posterior, 166, 178
  nerve, anterior, 179
    inferior, 170
    posterior, 179
Descendens noni nerve, 152, 158
Depressor epiglottidis, 211
Digastric muscle, 144
  nerve, 152
Digital arteries, radial, 82, 89
  ulnar, 82
  nerves of median, 84

Digital nerves of radial, 70
  of ulnar, 84
Dissection of arm, back, 54
  front, 37
  axilla, 1
  axillary vessels, 14
  base of skull, 104
  bend of elbow, 27
  brachial plexus, 14
  carotid artery, common, 141
    external, 141
    internal, 180
  cranial nerves in neck, 181
  forearm, back, 91, 98
    front, 63, 71
  hand, back, 91
    palm, 79, 86
  larynx, 201
  neck, back, 159
    front, 121
  neck, anterior triangle, 141
    posterior triangle, 121
  nose, 215
  orbit, 110, 117
  pharynx, 190
  pterygoid region, 156, 168
  scapula muscles, 48
  shoulder, 48
  soft palate, 192
  subclavian artery, 153
  submaxillary region, 172
  superior maxillary nerve, 177
  upper limb, 1
Dorsal artery of tongue, 175
  of scapula, 23, 53
Ductus ad nasum, 219
  Stenonis, 138
  Whartonii, 174
  Riviniani, 174
Dura mater of skull, 105, 116
  nerves of, 106
  vessels of, 105

EIGHTH cranial nerve, 108
Elbow in dislocation, 55, 65
Eleventh cranial nerve, 108, 134, 187
Epiglottis, 204
  articulation, 205

Epiglottis, use, 196, 204
Ethmoidal arteries, 113
Eustachian tube, 194, 217
Extensor carpi radialis brevis, 93
　　　　　longus, 93
　　carpi ulnaris, 94
　　digiti minimi, 94
　　digitorum communis, 93
　　indicis, 100
　　ossis metacarpi, 99
　　primi internodii pollicis, 99
　　secundi internodii pollicis, 99
External cutaneous nerve of arm, 36
　　mammary artery, 8
Eye, arteries, 113
　　muscles, 111, 118
　　nerves, 115, 119
　　veins, 113

FACIAL artery, 150, 166, 183
　　nerve, 108, 151, 180
　　vein, 166
Falx cerebelli, 105
　　cerebri, 105
Fascia, brachial, 27, 38
　　cervical, 137
　　costo-coracoid, 19
　　forearm, 27
Fat in axilla, 13
　　in hollow of elbow, 67
Fifth cranial nerve, 107
First cranial nerve, 107, 221
Flexor minimi digiti, 87
　　carpi radialis, 64
　　　　ulnaris, 64
　　digitorum profundus, 72, 80
　　　　　　sublimis, 65, 79
　　pollicis longus, 71
　　　　　brevis, 87
Forearm, dissection of, 63
　　front, deep, 71
　　　　superficial, 63
　　back, deep, 98
　　　　superficial, 91
Fossæ of base of skull, 104
Fourth cranial nerve, 107, 114
Fracture of clavicle, 15
Frontal artery, 113
　　nerve, 115

GANGLIA cervical, 187

Ganglion, Gasserian, 107
　　lenticular, 110
　　ophthalmic, 119
　　submaxillary, 176
Genio-hyo-glossus, 173
　　hyoideus, 172
Gland lachrymal, 111
　　parotid, 138
　　sub-lingual, 139, 174
　　submaxillary, 139
Glands, axillary, 12, 26
　　cervical, 143
　　submaxillary, 139, 174
Glosso-pharyngeal nerve, 108, 175, 185
Glottis, 207
Gustatory artery, 166, 176
　　nerve, 170, 176

HAND, dissection of, 78
　　back, 91
　　palm, 78, 86
Hollow before elbow, 66
Humerus, fracture of, 52
Hyo-glossus muscle, 173
Hyoid bone, 201
Hypoglossal nerve, 109, 152, 175, 187

INDICATOR muscle, 100
Inferior maxillary nerve, 169
Infra-orbital artery, 178
　　nerve, 180
　　trochlear nerve, 115
Infra-spinatus muscle, 50
Innominate artery, 154
　　vein, 156
Intercostal artery, superior, 156
　　cutaneous nerves, 11
Intercosto-humeral nerve, 12, 25
Internal cutaneous of arm, 25, 36
Interosseous arteries of hand, 89
　　artery, anterior, 76, 101
　　　　posterior, 100
　　muscles of hand, 88
　　nerve, anterior, 77
　　　　posterior, 102
Isthmus faucium, 195
　　of thyroid body, 214

JACOBSON'S nerve, 185
Jugular vein anterior, 140, 157
　　external, 131, 151, 157
　　internal, 151, 157

LABIAL artery, inferior, 166
Lachrymal artery, 113
    duct, 219
    gland, 111
    nerve, 114
Large artery of thumb, 89
Laryngeal arteries, 192, 213
    nerve, external, 186, 193
        inferior, 186, 212
        superior, 186, 211
    pouch, 207
Larynx, 206
    aperture, 196
    articulations, 204
    cartilages, 202
    interior, 206
    muscles, 209
    mucous membrane, 208
    nerves, 211
    ventricle, 207
    vessels, 213
Lateral cutaneous nerves of thorax, 11
Latissimus dorsi, 2, 17, 49
Lateral sinus, 117
Lenticular ganglion, 119
Levator anguli oris, 177
    scapulæ, 49, 123
    labii superioris, 177
    palati, 198
    palpebræ superioris, 112
    pharyngis, 191
Ligamenta brevia, 80
Ligaments of the larynx, 204
Ligamentum stylo-maxillare, 171
                  hyoidean, 173
Ligature of arteries
    axillary, 6, 21
    brachial, 34, 42
    carotid, common, 147
        external, 149
        internal, 148
    lingual, 150
    radial, 68
    subclavian, third part, 129
                second part, 156
    ulnar, 75
Limb, upper, dissection of, 1
Lingual artery, 150, 175
    vein, 175
Longus colli muscle, 182
Lumbricales of hand, 88
Lymphatic duct, left, 157
            right, 157

Lymphatics of arm, 47
    axilla, 12
    neck, 135

MAMMARY artery, external, 8
    internal, 156
Masseter muscle, 164
Masseteric artery, 166
    nerve, 169
Maxillary artery, internal, 150, 165
    nerve, inferior, 167
        superior, 170
Meatuses of the nose, 218
Median basilic vein, 29
    cephalic vein, 29
    nerve, 47, 70, 84
    veins, 28
Meningeal vessels, 109, 166, 169
    nerves, 106
    veins, 110
Metacarpal artery, 76
Motor oculi nerve, 107, 114, 120
Movement of radius, 72
Musculo-cutaneous nerve, 25
    spiral nerve, 46, 61, 102
Musculus abductor minimi digiti, 87
        indicis, 88
        pollicis, 86
    adductor minimi digiti, 87
        pollicis, 86
    anconeus, 94
    arytænoideus, 210
    arytæno-epiglottideus infer., 211
    azygos uvulæ, 199
    biceps humeralis, 18, 39
    brachialis anticus, 39
    buccinator, 164
    circumflexus palati, 198
    complexus, 159
    constrictor inferior, 191
        isthmi faucium, 199
        medius, 191
        superior, 191
    coraco-brachialis, 3, 18
    crico-arytænoideus lateralis, 209
                posticus, 209
    crico-thyroideus, 210
    deltoides, 51
    depressor epiglottidis, 211

Musculus digastricus, 144
  extensor carpi radialis longus, 93
  extensor carpi radialis brevis, 93
  extensor carpi ulnaris, 94
    minimi digiti, 94
    digitorum manus, 93
    indicis, 100
    ossis metacarpi pollicis, 59
    primi internodii pollicis, 99
    secundi internodii pollicis, 59
  flexor brevis minimi digiti, 87
    carpi radialis, 64
    ulnaris, 64
    pollicis brevis, 87
    longus, 71
    profundus digit., 72, 80
    sublimis digit., 65, 79
  genio-hyo-glossus, 173
    hyoideus, 172
  hyo-glossus, 173
  indicator, 100
  infra-spinatus, 50
  interossei manus, 88
  latissimus dorsi, 2, 17, 49
  levator anguli oris, 177
    scapulæ, 49, 123
    labii superioris, 177
    alæ nasi, 178
    palati, 198
    palpebræ, 112
    pharyngis externus, 191
      internus, 192
    uvulæ, 199
  longus colli, 182
  lumbricales manus, 88
  massetericus, 164
  mylo-hyoideus, 145
  obliquus capitis infer., 160
    super., 160
  oculi inferior, 118
    superior, 111

Musculus omo-hyoideus, 123, 144, 154
  opponens minimi digiti, 87
    pollicis, 86
  orbicularis palpebrarum, 177
  palato-glossus, 199
    pharyngeus, 199
  palmaris brevis, 79
    longus, 64
  pectoralis major, 2, 16
    minor, 2, 16
  pronator quadratus, 72
    radii teres, 63
  pterygoideus externus, 164
    internus, 164
  rectus capitis anticus major, 181
    minor, 182
    lateralis, 181
    posticus major, 160
      minor, 160
  oculi externus, 112, 118
    inferior, 118
    internus, 118
    superior, 112
  rhomboideus, major, 49
    minor, 49
  salpingo-pharyngeus, 192
  scalenus anticus, 123, 153
    medius, 124
    posticus, 124
  semi spinalis colli, 160
  serratus magnus, 3, 17, 124
  splenius capitis, 123
  sterno-cleido-mastoideus, 122, 137
    hyoideus, 144
    thyroideus, 144
  stylo-glossus, 173
    hyoideus, 144
    pharyngeus, 191
  subclavius, 17, 154
  subscapularis, 3, 17
  supinator radii brevis, 98
    longus, 59, 65, 92
  supra-spinatus, 50
  temporalis, 163
  tensor palati, 198
  teres major, 2, 18, 49
    minor, 50

INDEX. 229

Musculus thyro-arytænoideus, 210
    hyoideus, 144
    trapezius, 122
    triceps extensor humeri, 39, 51, 54, 59
Mylo-hyoid artery, 166
    muscle, 145
    nerve, 152

NARES, posterior, 195
Nasal artery, 113
    cartilages, 216
    duct, 219
    fossæ, 216
    nerve, 115, 221
Neck, anterior triangle, 141
    posterior triangle, 124
    dissection of, 121
Nerve of latissimus, 25
    levator anguli scapulæ, 189
    pterygoid muscles, 170
    rhomboid muscles, 134
    serratus magnus, 25, 134
    subclavius, 134, 158
    teres major, 25
    minor, 53
Nervus abducens oculi, 108, 120
    accessorius spinalis, 108, 134, 187
    auditorius, 108
    auricularis magnus, 133, 152
        posterior, 134, 152
    auriculo-temporalis, 152, 170
    buccalis, 169
    buccinatorius, 169
    cardiaci, 186
    cardiacus infer., 188
        medius, 188
        super., 188
    cervicales, 162, 189
    cervicalis superficialis, 133, 141
    ciliares, 115, 119
    circumflexus, 11, 53
    cutaneous externus brachii, 25, 36
        internus major, 25, 36
        minor, 25, 36
        palmaris, 77
    dentalis anterior, 179
        inferior, 170
        posterior, 179
    decendens noni, 152, 158

Nervus diaphragmaticus, 158
    digastricus, 152
    digitalis, median, 84
        radial, 70
        ulnar, 85
    dorsales ulnaris, 85
    facialis, 108, 151, 180
    frontalis, 115
    glosso-pharyngeus, 108, 175, 185
    gustatorius, 170, 176
    hyoglossus, 109, 152, 175, 187
    infra-orbitalis, 180
        maxillaris, 140
        trochlearis, 115
    intercosto-cutanei, 11
    interosseus anticus, 77
        posticus, 102
    lachrymalis, 114
    laryngeus externus, 186, 193
        inferior, 186, 212
        superior, 186, 211
    massetericus, 182
    maxillaris inferior, 168
        superior, 179
    medianus, 47, 70, 77, 84
    meningealis, 106
    motor oculi, 107, 114, 120
    musculo-cutaneous brachii, 25, 36
        spiralis, 46, 61, 102
    mylo-hyoideus, 152
    nasalis, 115, 221
    occipitalis major, 162
        minor, 133, 162
    olfactorius, 107, 221
    ophthalmicus, 114
    opticus, 107, 119
    orbitalis, 179
    palmaris cutaneus med., 70
        ulnar, 77
    perforans Casserii, 25, 36
    petrosis magnus, 109
    pharyngeus, 186
    phrenicus, 158, 189
    pneumo-gastricus, 108, 158, 185
    radialis, 70
    recurrens vagi, 186, 212
    spheno-palatinus, 179, 221
    stylo-hyoideus, 152
    suboccipitalis ram. ant., 189
        post., 162
    subscapularis, 25
    supra-orbitalis, 114

230 INDEX.

Nervus supra-scapularis, 134, 158
    trochlearis, 115
    sympatheticus cervicis, 187
    temporalis profundus, 169
    temporo-facialis, 151
    thoracici anteriores, 25
    thoracicus posterior, 25, 134
    thyro-hyoideus, 152
    trigeminus, 107
    trochlearis, 107, 114
    tympanicus, 185
    ulnaris, 47, 77, 85, 90
Ninth cranial nerve, 108, 185
Nose cartilages, 216
    cavity, 216
    meatuses, 218
    mucous membrane, 219
    nerves, 220
    vessels, 220
Nostril, 216
Nutritious artery, humeral, 45

OBLIQUUS capitis, inferior, 160
    superior, 160
    oculi, inferior, 118
    superior, 111
Occipital artery, 150, 161
    sinus, 117
    nerve, large, 162
    small, 133, 162
Œsophagus, aperture of, 196
Olecranon, fracture of, 55, 65
Olfactory nerve, 107, 220
    region, 220
Omo-hyoid muscle, 123, 144, 154
Ophthalmic artery, 113
    ganglion, 119
    nerve, 114
    vein, 113
Opponens pollicis muscle, 86
Optic commissure, 106
    nerve, 107, 119
Orbicularis palpebrarum, 177
Orbit, dissection of, 110, 117
    muscles, 111, 117
    nerves, 114
    vessels, 112
Orbital branch of nerve, 120, 179
Os hyoides, 201

PALATE, soft, 197
    use of, 200
Palatine arteries, 183

Palato-glossus, 199
    pharyngeus, 199
Palm of the hand, 78
    dissection, 78
    cutaneous nerves, 78
Palmar arch, deep, 89
    superficial, 81
Palmaris brevis, 79
    longus, 64
Parotid gland, 138
Pectoralis major, 2, 16
    minor, 2, 16
Peculiarities in arteries
    axillary, 6
    brachial, 35, 44
    radial in forearm, 32, 68, 96
    subclavian, 129
    ulnar in forearm, 32
    in palm, 83
Perforating arteries, interosseous, 100
    palmar, 89
Perforans Casserii nerve, 25, 36
Petrosal nerve, large, 109
    sinuses, 116
Pharynx, dissection, 190
    interior, 193
    muscles, 190
    openings, 194
Pharyngeal ascending artery, 183
    nerve, 186
Phrenic nerve, 158, 189
Pillars of soft palate, 197
Pituitary membrane, 219
Plate
  1. The axilla
  2. The axillary vessels
  3. The cutaneous vessels of forearm
  4. The brachial vessels
  5. The shoulder and scapula
  6. The arm, back
  7. The spiral nerve and vessels
  8. The forearm, front
  9. The forearm, deep view
  10. The palm of hand, superficial and deep views
  11. The forearm, back
  12. The forearm, deep view
  13. The base of skull and orbit, superficial view
  14. The base of skull and orbit, deep view
  15. The neck, posterior triangle
  16. The neck, surface view

Plate
17. The neck, anterior triangle
18. The subclavian vessels
19. The neck, view behind
20. The pterygoid region
21. The pterygoid, deep view
22. The submaxillary region
23. The upper maxillary nerve
24. The internal carotid artery
25. The pharynx, surface view
26. The pharynx, interior
27. The larynx and vocal apparatus
28. The nose cavity
Platysma myoides, 121
Plexus brachial, 25, 158, 189
    cervical, 133, 189
    pharyngeal, 186
Plugging the nares, 217
Pneumo-gastric nerve, 108, 158, 185
Pomum Adami, 202
Portio dura, 108
    mollis, 108
Posterior triangle of neck, 124
Pouch laryngeal, 207
Profunda artery, arm, 45
    neck, 161
Pronator quadratus, 72
    radii teres, 63
Pterygoid arteries, 166
    nerves, 170, 171
Pterygoideus externus, 164
    internus, 164
Pterygo-maxillary region, 163

RADIAL artery, 68, 88, 96
    nerve, 70
    veins, 68
Radius, fracture, 73
    movement, 72
Ranine artery, 175
Rectus capitis anticus major, 181
    minor, 182
    posticus major, 160
    minor, 160
    lateralis, 181
    oculi externus, 112, 118
    inferior, 118
    internus, 118
    superior, 112
Recurrent interosseous artery, 101
    radial, 69, 101
    ulnar anterior, 76

Recurrent ulnar posterior, 76
    nerve, 186, 212
Rhomboideus major, 49
    minor, 50
Rima glottidis, 207

SACCULUS laryngis, 207
Salpingo-pharyngeus, 192
Scaleni muscles, 123, 153
Scapular artery, posterior, 130
    muscles, 17, 48
Schneiderian membrane, 219
Second cranial nerve, 107, 119
Semi-spinalis colli, 160
Septum nasi, 216
Serratus magnus muscle, 3, 17, 125
Seventh cranial nerve, 108
Sheath, axillary, 10
    digital of fingers, 80
Sinuses of the skull, 116
Sixth cranial nerve, 108
Soft palate, 197
    use of, 200
Spheno-palatine nerves, 179, 221
Spinal accessory nerve, 108, 134, 187
Splenius capitis, 122
Spongy bones, 218
Stenon's duct, 138
Sterno-cleido-mastoideus, 122, 137
    hyoideus, 144
    thyroideus, 144
Straight sinus, 117
Stylo-hyoid ligament, 173
    hyoideus, 144
    glossus, 173
    maxillary ligament, 171
    pharyngeus, 191
Subanconeous muscle, 59
Subclavian artery, 127, 154
    vein, 157
Subclavius muscle, 17, 154
Sublingual artery, 175
    gland, 139, 174
Submaxillary ganglion, 176
    gland, 139, 174
    region, 172
Submental vessels, 150
Suboccipital nerve, ant. branch, 189
    post. branch, 162
Subscapular nerve, 25
    vessels, 8
Subscapularis muscle, 3, 17

INDEX.

Superficial cervical artery, 131
    nerve, 133, 141
    volar artery, 69
Supinator radii brevis, 98
    longus, 59, 65, 92
Supra-orbital artery, 113
    nerve, 114
    scapular artery, 130, 156
      nerve, 134, 158
      vein, 157
    spinatus muscle, 50
    trochlear nerve, 115
Spermatic cord, cervical, 187

TEMPORAL arteries, 150, 166
    muscle, 163
    nerve, 169
Temporo-facial nerve, 152
Tensor palati, 198
Tenth cranial nerve, 108, 185
Tentorium cerebelli, 105
Teres major muscle, 2, 49
    minor muscle, 50
Third cranial nerve, 107, 114, 119
Thoracic nerve, posterior, 134
    acromial artery, 23
    alar, 8, 23
    humeral, 23
    long, 8, 23
    superior, 23
Thyro-arytænoid articulations, 206
    muscle, 210
    epiglottid ligament, 205
    hyoid membrane, 205
      muscle, 144
      nerve, 152
Thyroid artery, inferior, 156, 182, 214
    superior, 150, 183, 214
    axis, 156
    body, 214
    cartilage, 202
    veins, 214
Tonsil, 197
    artery of, 183
Torcular Herophili, 117
Trachea, connections, 215
    structure, 215
Tracheal glands, 215
Transverse cervical artery, 131, 156
    vein, 157
    sinus, 116
Trapezius muscle, 122

Triangle of neck, anterior, 141
    posterior, 124
Triangular cartilage, nasal, 217
Triceps extensor cubiti, 51, 54, 59
Trigeminal cranial nerve, 107
Trochlea, 111
Trochlear nerve, infra, 115
    supra, 115
Turbinate bones, 218
Twelfth cranial nerve, 109, 175, 187
Tympanic artery, 183

ULNAR artery, 69, 74
    nerve, 47, 77, 85, 90
    veins, deep, 75
      superficial, 28

VAGUS cranial nerve, 108, 185
Vein, alveolar, 167
    auricular posterior, 151
    axillary, 9
    basilic, 29
    brachial, 40
    cephalic, 24, 30
    deep cervical, 161
    facial, 166
      deep, 167
    innominate, 157
    jugular anterior, 149, 157
      external, 131, 149, 151
      internal, 151, 157
    laryngeal, 213
    lingual, 175
    median of forearm, 28
      basilic, 29
      cephalic, 29
    maxillary internal, 151, 167
    nasal, 220
    ophthalmic, 113
    pterygoid, 167
    radial cutaneous, 29
      deep, 68
    subclavian, 156
    temporal, 151
    thyroid, 214
    ulnar cutaneous, 28
      deep, 75
    vertebral, 155, 161
Vena cava superior, 157
Ventricle of larynx, 207
Vertebral artery, 109, 156, 161
    vein, 155, 161

Vessels of dura mater, 105
Vocal cords, 206
    use of, 207

WOUNDS of arteries,
    brachial, 32, 34
    palmar arch, superficial, 82

Wounds of arteries,
    palmar arch, deep. 90
    radial, 68
    ulnar in forearm, 76
Wharton's duct, 174
Wrisberg's nerve, 36, 46
    cartilages, 204

www.ingramcontent.com/pod-product-compliance
Lightning Source LLC
Chambersburg PA
CBHW031329230426
43670CB00006B/288